Springer Tracts in Advanced Robotics 78

Editors

Prof. Bruno Siciliano
Dipartimento di Informatica
e Sistemistica
Università di Napoli Federico II
Via Claudio 21, 80125 Napoli
Italy
E-mail: siciliano@unina.it

Prof. Oussama Khatib
Artificial Intelligence Laboratory
Department of Computer Science
Stanford University
Stanford, CA 94305-9010
USA
E-mail: khatib@cs.stanford.edu

For further volumes:
http://www.springer.com/series/5208

Editorial Advisory Board

Oliver Brock, TU Berlin, Germany
Herman Bruyninckx, KU Leuven, Belgium
Raja Chatila, LAAS, France
Henrik Christensen, Georgia Tech, USA
Peter Corke, Queensland Univ. Technology, Australia
Paolo Dario, Scuola S. Anna Pisa, Italy
Rüdiger Dillmann, Univ. Karlsruhe, Germany
Ken Goldberg, UC Berkeley, USA
John Hollerbach, Univ. Utah, USA
Makoto Kaneko, Osaka Univ., Japan
Lydia Kavraki, Rice Univ., USA
Vijay Kumar, Univ. Pennsylvania, USA
Sukhan Lee, Sungkyunkwan Univ., Korea
Frank Park, Seoul National Univ., Korea
Tim Salcudean, Univ. British Columbia, Canada
Roland Siegwart, ETH Zurich, Switzerland
Gaurav Sukhatme, Univ. Southern California, USA
Sebastian Thrun, Stanford Univ., USA
Yangsheng Xu, Chinese Univ. Hong Kong, PRC
Shin'ichi Yuta, Tsukuba Univ., Japan

STAR (Springer Tracts in Advanced Robotics) has been promoted under the auspices of EURON (European Robotics Research Network)

Tin Lun Lam and Yangsheng Xu

Tree Climbing Robot

Design, Kinematics and Motion Planning

Authors

Tin Lun Lam
The Chinese University of Hong Kong
William M.W. Mong Engineering Building
RM 302
Shatin, N.T.
Hong Kong
China, People's Republic
E-mail: tllam@mae.cuhk.edu.hk

Yangsheng Xu
The Chinese University of Hong Kong
University Administration Building
RM 212
Shatin, N.T.
Hong Kong
China, People's Republic
E-mail: ysxu@cuhk.edu.hk

ISSN 1610-7438   e-ISSN 1610-742X
ISBN 978-3-642-28310-9   e-ISBN 978-3-642-28311-6
DOI 10.1007/978-3-642-28311-6
Springer Heidelberg New York Dordrecht London

Library of Congress Control Number: 2012933103

© Springer-Verlag Berlin Heidelberg 2012

This work is subject to copyright. All rights are reserved by the Publisher, whether the whole or part of the material is concerned, specifically the rights of translation, reprinting, reuse of illustrations, recitation, broadcasting, reproduction on microfilms or in any other physical way, and transmission or information storage and retrieval, electronic adaptation, computer software, or by similar or dissimilar methodology now known or hereafter developed. Exempted from this legal reservation are brief excerpts in connection with reviews or scholarly analysis or material supplied specifically for the purpose of being entered and executed on a computer system, for exclusive use by the purchaser of the work. Duplication of this publication or parts thereof is permitted only under the provisions of the Copyright Law of the Publisher's location, in its current version, and permission for use must always be obtained from Springer. Permissions for use may be obtained through RightsLink at the Copyright Clearance Center. Violations are liable to prosecution under the respective Copyright Law.

The use of general descriptive names, registered names, trademarks, service marks, etc. in this publication does not imply, even in the absence of a specific statement, that such names are exempt from the relevant protective laws and regulations and therefore free for general use.

While the advice and information in this book are believed to be true and accurate at the date of publication, neither the authors nor the editors nor the publisher can accept any legal responsibility for any errors or omissions that may be made. The publisher makes no warranty, express or implied, with respect to the material contained herein.

Printed on acid-free paper

Springer is part of Springer Science+Business Media (www.springer.com)

To our families

Foreword

Robotics is undergoing a major transformation in scope and dimension. From a largely dominant industrial focus, robotics is rapidly expanding into human environments and vigorously engaged in its new challenges. Interacting with, assisting, serving, and exploring with humans, the emerging robots will increasingly touch people and their lives.

Beyond its impact on physical robots, the body of knowledge robotics has produced is revealing a much wider range of applications reaching across diverse research areas and scientific disciplines, such as: biomechanics, haptics, neurosciences, virtual simulation, animation, surgery, and sensor networks among others. In return, the challenges of the new emerging areas are proving an abundant source of stimulation and insights for the field of robotics. It is indeed at the intersection of disciplines that the most striking advances happen.

The *Springer Tracts in Advanced Robotics (STAR)* is devoted to bringing to the research community the latest advances in the robotics field on the basis of their significance and quality. Through a wide and timely dissemination of critical research developments in robotics, our objective with this series is to promote more exchanges and collaborations among the researchers in the community and contribute to further advancements in this rapidly growing field.

The monograph by Tin Lun Lam and Yangsheng Xu is based on the first author's doctoral thesis under the supervision of his co-author. Tree-climbing robots have been receiving an increasing interest in the research community in view of the number of challenges posed to the design by the application scenario. Several approaches in autonomous tree-climbing, including the sensing methodology, cognition of the environment, path planning and motion planning are proposed in the text. Further, a novel biologically inspired prototype is presented and its enhanced performance over the state of the art in the field is demonstrated in a number of experiments for both known and unknown environments.

The first contribution to the series on climbing robots, this volume constitutes a fine addition to STAR!

Naples, Italy Bruno Siciliano
December 2011 STAR Editor

Preface

Climbing robot is a challenging research topic that has gained much attention from researchers. Most of the climbing robots reported in the literature are designed to work on manmade structures, such as vertical walls, glass windows or structural frames. There are seldom robots designed for climbing natural structures such as trees. Trees and manmade structures are very different in nature. For example, tree surfaces are seldom flat and smooth, and some trees have soft bark that peels off easily. It brings different aspects of technical challenges to the robot design. In the state-of-the-art tree-climbing robots, the workspaces are restricted on tree trunks only. They cannot act like arboreal animals such as squirrels to reach any position on irregularly shaped trees with branches. As branches and curvature are presented in many kinds of trees, the application of these robots is strongly restricted. It is clearly that the tree-climbing technology in robotics still has big room for improvement.

Through billions of years of evolution, many types of arboreal animals have evolved and developed diverse methods to deal with these challenges. The rigorous competition of natural selection process confirms the effectiveness and efficiency of the present solutions in nature. It is believed that the solution in nature can inspire an idea to solve the captioned technical challenges in certain level.

In this book, a comprehensive study and analysis of both natural and artificial tree-climbing methods is presented. It provides a valuable reference for robot designers to select appropriate climbing methods in designing tree-climbing robots for specific purposes. Based on the study, a novel bio-inspired tree-climbing robot with several breakthrough performances has been developed and presents in this book. It is capable of performing various actions that is impossible in the state-of-the-art tree-climbing robots, such as moving between trunk and branches. This book also proposes several approaches in autonomous tree-climbing, including the sensing methodology, cognition of the environment, path planning and motion planning on both known and unknown environment.

This book originates from the PhD thesis of the first author at the Chinese University of Hong Kong, supervised by the second author. In this book, you can find a collection of the cutting edge technologies in the field of tree-climbing robot and the ways that animals climb. You can also find the development and application of a novel type of climbing mechanism. Although the novel mechanism is applied for tree climbing in this book, it has high potential to apply on others fields due to its distinguish characteristics. In addition, the work also illustrates a successful example of biomimetics as several important aspects in the work such as maneuver mechanism and the method of environment cognition in autonomous control.

This book is appropriate for postgraduate students, research scientists and engineers with interests in climbing robots and biologically inspired robots. In particular, the book will be a valuable reference for those interested in the topics of mechanical design, implementation, and autonomous control for tree-climbing robots.

The Chinese University of Hong Kong　　　　　　　　　　　　　*Tin Lun Lam*
November 2011　　　　　　　　　　　　　　　　　　　　　　　*Yangsheng Xu*

Acknowledgements

This book would not have been possible without the support of many people. First and foremost, we would like to acknowledge the members of the Advanced Robotics Laboratory at the Chinese University of Hong Kong and especially grateful to Dr. Huihuan Qian, Mr. Wing Kwong Chung, Mr. Kai Wing Hou, Mr. Yongquan Chen, and Ms. Hoi Yee Nam for their invaluable assistance, support, and enlightening discussions. We would also like to acknowledge Mr. Allan Mok for his technical support and guidance. Beside, we would also like to show our gratitude to Prof. Lilong Cai, Prof. Yun-hui Liu, and Prof. Changling Wang, who given invaluable suggestion to the works presented in this book. We greatly appreciate them taking the time to evaluate the works.

The first author would like to particularly express his sincere gratitude to his supervisor, Prof. Yangsheng Xu who originally suggested the research topic to him and gave continuous encouragement and invaluable guidance. Special thanks also to all his family members and friends for their help and continuous encouragement. The deepest gratitude of the first author goes to his parents for their unconditional love and support throughout his life, and encouraging him to pursue his interests. This book would simply not have been possible without them.

The second author would like to thank Prof. Richard Paul who taught him on robotics and shared vast wealth of engineering knowledge and philosophy of life to him. The second author would also like to express his gratitude to his family and especially his wife Nancy and his son Peter for their moral support and encouragement during the course of study. The second author thanks them for always being there when he needed them.

Contents

1	**Introduction**	1
	1.1 Background	1
	1.2 Motivation	2
	1.3 Outline of the Book	3
2	**State-of-the-Art Tree-Climbing Robots**	5
	2.1 WOODY	5
	2.2 Kawasaki's Pruning Robot	6
	2.3 Seirei Industry's Pruning Robot	6
	2.4 TREPA	7
	2.5 RiSE	7
	2.6 Modular Snake Robot	8
3	**Methodology of Tree Climbing**	9
	3.1 Tree-Climbing Methods in Nature	9
	3.2 Artificial Tree-Climbing Methods	11
	3.3 Design Principles for Tree-climbing Robots	14
	3.4 Ranking of the Tree-Climbing Methods	15
	3.4.1 Maneuverability	15
	3.4.2 Robustness	17
	3.4.3 Complexity	17
	3.4.4 Adaptability	18
	3.4.5 Size	19
	3.4.6 Speed	19
	3.5 Summary	20
4	**A Novel Tree-Climbing Robot: Treebot**	23
	4.1 Objectives	23
	4.2 Approach to the Robot Design	24

4.3	Structure		24
	4.3.1	Tree Gripper	26
	4.3.2	Continuum Body	27
	4.3.3	Semi-passive Joint	32
	4.3.4	Sensors	34
4.4	Locomotion		34
4.5	Hardware Prototype		37
4.6	Energy Consumption		40
4.7	Accessories		40
4.8	Control		41
	4.8.1	Control Architecture	41
	4.8.2	Manual Control	43
4.9	Experiments		43
	4.9.1	Generality	44
	4.9.2	Transition Motion	44
	4.9.3	Turning Motion	45
	4.9.4	Slope Climbing	46
	4.9.5	Payload	50
4.10	Performance Comparison		50
4.11	Summary		50

5 Optimization of the Fastening Force ... 55
  5.1 Gripping Configuration ... 55
  5.2 Generation of the Adhesive Force ... 57
  5.3 Optimization of the Spine Installation Angle ... 60
  5.4 Generation of the Directional Penetration Force ... 66
  5.5 Experimental Results ... 69
  5.6 Summary ... 72

6 Kinematics and Workspace Analysis ... 73
  6.1 Kinematic Analysis ... 73
    6.1.1 Configuration of Treebot ... 73
    6.1.2 Kinematics of the Continuum Body ... 73
    6.1.3 Kinematics of Treebot ... 76
    6.1.4 Tree Model ... 78
  6.2 Workspace Analysis ... 80
    6.2.1 Physical Constraints ... 81
    6.2.2 Admissible Workspace on a Tree Surface ... 85
  6.3 Summary ... 89

7 Autonomous Climbing ... 91
  7.1 Autonomous Climbing Strategy ... 92
  7.2 Tree Shape Approximation ... 92
    7.2.1 Exploring Strategy ... 92
    7.2.2 Arc Fitting ... 94

		7.2.3	Tree Shape Reconstruction	98
		7.2.4	Angle of Change to the Upper Apex	100
		7.2.5	Tree Radius Approximation	100
	7.3	Motion Planning		102
		7.3.1	Strategy 1	102
		7.3.2	Strategy 2	104
		7.3.3	Strategy 3	106
		7.3.4	Verification of Target Position	108
	7.4	Experiments		108
		7.4.1	Tree Shape Approximation	109
		7.4.2	Motion Planning	110
		7.4.3	Climbing a Tree with Branches	113
	7.5	Summary		113
8	**Global Path and Motion Planning**			**117**
	8.1	State Space Formulation		118
	8.2	Path Planning		120
		8.2.1	Dynamic Programming	120
		8.2.2	Dynamic Environment	122
	8.3	Motion Planning		122
		8.3.1	Strategy of Motion Planning	123
		8.3.2	Posture of the Robot	126
		8.3.3	Adaptive Path Segmentation	127
	8.4	Simulations		128
		8.4.1	Global Path Planning	130
		8.4.2	Motion Planning	130
	8.5	Experiments		136
	8.6	Summary		137
9	**Conclusions and Future Work**			**139**
	9.1	Conclusion		139
		9.1.1	Methodology and Design Principle for Tree-climbing Robots	139
		9.1.2	A Novel Tree-climbing Robot with Distinguish Performance	139
		9.1.3	Kinematics and Workspace Analysis	140
		9.1.4	Autonomous Climbing Strategy in an Unknown Environment	141
		9.1.5	Global Path and Motion Planning on Climbing Irregularly Shaped Trees	141
	9.2	Future Research Directions		142
		9.2.1	Fastening Mechanism	142
		9.2.2	Continuum Mechanism	142
		9.2.3	Map Building and Localization	143

A Derivation of Equations ... 145
A.1 Kinematics of the Continuum Body ... 145
A.1.1 Inverse Kinematics ... 145
A.1.2 Forward Kinematics ... 146
A.1.3 Mapping between the Posture and the Cartesian Coordinate ... 148
A.2 Kinematics of Treebot ... 151
A.2.1 Mapping between the Posture and the Cartesian Coordinate ... 151

References ... 155

Index ... 161

List of Figures

3.1 Illustration of the gripping angles by using different methods. (a) Wet adhesion or Van der Waals force; (b) Claw gripping; (c) Interlock method. 10
3.2 Climbing gaits of different maneuvering methods. (a) Extend-contract gait; (b) Quadrupedal gait; (c) Tripod gait. 20
3.3 Performance evaluation of the existing tree-climbing robots based on the design principles. 21
4.1 Expected performance of a tree-climbing robot with the use of claw penetration method for fastening and extend-contract with bend method for maneuvering. 24
4.2 Structure of Treebot. 25
4.3 Structure of the tree gripper. (a) Isometric view; (b) Top view; (c) Cross-sectional view. 27
4.4 Gripping mechanism of the tree gripper in cross-sectional view. 28
4.5 Prototype of the tree gripper. 29
4.6 Mechanical design of the proposed continuum body. 31
4.7 Prototype of the continuum body and illustration of the admissible motions: (a) contraction; (b) extension; (c) forward bending; (d) right bending; (e) left bending. 32
4.8 Compliance of the continuum body: (a) shearing and (b) twisting. 33
4.9 Design of the semi-passive joint. 34
4.10 Locking mechanism of the semi-passive joint. 35
4.11 Hardware prototype and the installation position of the semi-passive joint. (a) Initial orientation; (b) Twisted about the z-axis. 36
4.12 Semi-passive joint: (a) unlocked; (b) locked. 36
4.13 A complete climbing gait of Treebot (moving forward). 36
4.14 Motions to avoid an obstacle. 37
4.15 The first prototype: Treebot. 38
4.16 The second prototype: Treebot-Auto. 39
4.17 A camera module. 41

4.18	A photovoltaic module.	41
4.19	Control architecture of Treebot.	42
4.20	Interface of the ground station.	42
4.21	Configuration of the control panel to control Treebot.	44
4.22	Climbing test on different types of trees: (a) Callistemon viminalis; (b) Bambusa ventricosa; (c) Araucaria heterophylla; (d) Cinnamomum camphora; (e) Bambusa vulgaris var. Striata; (f) Melaleuca quinquenervia.	46
4.23	Branch transition on a Bauhinia blakeana.	47
4.24	Branch transition on a Delonix regia.	48
4.25	Turning motion on a Bauhinia blakeana.	49
4.26	103 degrees slope climbing.	51
4.27	110 degrees slope climbing.	52
4.28	Climbs with 1.75kg payload.	54
5.1	Representation of the gripping position.	56
5.2	(a) Notations to find out the minimal radius of a substrate by interlocking method. (b) Notations to find out the minimal radius of a substrate that the gripper is capable of gripping the closer-half of the substrate only.	56
5.3	Notations for the gripping curvature and the gripper parameters.	58
5.4	Notations for defining the spine direction vector.	59
5.5	Relationships among the spine insert angle, adhesive force, and shear force [35].	61
5.6	Representative gripping orientations.	61
5.7	Relationships among the spine installation angle, the curvature of the gripping surface and the normalized pull-in force at $\sigma = 0$.	62
5.8	Relationships among the spine installation angle, the curvature of the gripping surface and the normalized pull-in force at $\sigma = \pi/8$.	62
5.9	Relationships among the spine installation angle, the curvature of the gripping surface and the normalized pull-in force at $\sigma = \pi/4$.	63
5.10	Relationships among the spine installation angle, the curvature of the gripping surface and the normalized pull-in force at $\sigma = 3\pi/8$.	63
5.11	Relationships among the spine installation angle, the curvature of the gripping surface and the normalized pull-in force at $\sigma = \pi/2$.	64
5.12	Normalized pull-in force generated by the gripping posture Orientation 1 in different spine installation angles and gripping curvatures.	64
5.13	Normalized pull-in force generated by the gripping posture Orientation 2 in different spine installation angles and gripping curvatures.	65
5.14	Normalized pull-in force generated by the gripping posture Orientation 3 in different spine installation angles and gripping curvatures.	65

List of Figures   XIX

5.15 Normalized pull-in force generated by the average of the gripping orientations in different spine installation angles and gripping curvatures. 66
5.16 Normalized pull-in force generated by different gripping orientations with optimal spine installation angle $\theta_s = 0.76$rad....... 67
5.17 Notations for the mechanism of the claw. 67
5.18 Free body diagrams of links AB and BC. 68
5.19 Experiments on different types of trees: (a) Bauhinia variegata var. candida; (b) Roystonea regia; (c) Taxodium distichum; (d) Cinnamomum camphora; (e) Khaya senegalensis; (f) Eucalyptus citriodora. 70

6.1 Configuration of Treebot. 74
6.2 Notations for defining the position and parameters of the continuum manipulator. 75
6.3 Notations for defining the kinematics of Treebot. 77
6.4 Relationship between the rear gripper and the tree model. 79
6.5 Reachable workspace of the continuum body. 81
6.6 Maximum length of extension of the continuum body with different bending curvature and bending direction at $l_{max} = 0.3m$. 82
6.7 Minimum length of extension of the continuum body with different bending curvature and bending direction at $l_{min} = 0m$. 83
6.8 Relationship between the limit of the climbing slope and the location of the center of mass. 84
6.9 Workspace of the gripper at each reachable position. 85
6.10 Relationship between the front gripper and the tree model. 87
6.11 Admissible gripping positions of the front gripper at $\theta_{rx} = 0$. 87
6.12 Admissible gripping positions of the front gripper at $\theta_{rx} = \pi/6$. 88
6.13 Admissible gripping positions of the front gripper at $\theta_{rx} = \pi/3$. 88
6.14 Admissible gripping positions of the front gripper at $\theta_{rx} = \pi/2$. 89

7.1 Flow chart of the autonomous climbing strategy. 93
7.2 Notations and procedures for arc fitting. (a) Path segment; (b) Transformation; (c) Plane fitting; (d) Arc fitting. 95
7.3 Tree shape approximation by the fitted arc. 99
7.4 The concept of finding an optimal angle of change. 101
7.5 Series of motions by Strategy 1. 103
7.6 Series of motions by Strategy 2. 105
7.7 Series of motions by Strategy 3. 107
7.8 Test 1: Tree shape approximation on a straight tree. (a) Approximation target and final exploring posture of Treebot; (b) Approximation result. 109
7.9 Test 2: Tree shape approximation on a straight tree. (a) Approximation target and final exploring posture of Treebot; (b) Approximation result. 110

7.10 Test 3: Tree shape approximation on a curved tree.
(a) Approximation target and final exploring posture of Treebot;
(b) Approximation result. 111
7.11 Experimental result of going to optimal path by Strategy 2. 112
7.12 Experimental result of going to optimal path by Strategy 3.
(a) Initial position; (b), (e) Exploring motion; (c), (f) Front gripper
gripping; (d), (g) Rear gripper gripping. 114
7.13 Experiment for climbing a tree with branches. (a) Initial position
in the first test; (b) Exploring posture in the first test; (c) Initial
position in the second test; (d) Exploring posture in the second
test. 115

8.1 Representation of the relationship among branches by using a tree
data structure. (a) Real tree structure; (b) Branch relationship as
represented by the tree data structure. 119
8.2 Tree surface discretization method. 119
8.3 State space representation to the path planning problem. 120
8.4 Coordinates and notations for the shape of the tree and the gravity
vector. 122
8.5 Procedures for 3D arc fitting: (a) A path segment;
(b) Transformation; (c) Plane fitting; (d) 2D arc fitting. 124
8.6 Optimal position and direction of the rear gripper. 125
8.7 The concept to determine the posture of the continuum body during
a contraction motion. 126
8.8 Procedures of the proposed motion planning strategy. 128
8.9 Experimental tree model. 129
8.10 Reward value of the selected state space. 130
8.11 Motion planning results by using Scheme 1. (a) Front view;
(b) Left view. 132
8.12 Motion planning results by using Scheme 2. (a) Front view;
(b) Left view. 133
8.13 Motion planning results by using Scheme 3. (a) Front view;
(b) Left view. 134
8.14 Motion planning results by using Scheme 4. (a) Front view;
(b) Left view. 135
8.15 (a) The target tree of climbing and the target climbing position;
(b) Approximated tree model and the solutions of path and motion
planning. 136
8.16 Climbing motions of Treebot according to the proposed path and
motion planning algorithm. 137

A.1 Notations of the continuum manipulator. 150
A.2 Notations of Treebot. 154

List of Tables

3.1	Summary of the tree-climbing methods applied in nature.	12
3.2	Summary of the artificial tree-climbing methods.	13
3.3	Ranking of the fastening and maneuvering methods in tree climbing (1=worst, 6=best).	16
4.1	Specifications of Treebot	40
4.2	Climbing performance on different species of trees	45
4.3	Comparison among the existing tree-climbing robots	53
5.1	Maximum pull-in force on different species of trees	71
7.1	Exploring strategy	94
8.1	Length, plane and arc fitness values of the path segments by Scheme 2.	131
8.2	Length, plane and arc fitness values of the path segments by Scheme 3.	133
8.3	Length, plane and arc fitness values of the path segments by Scheme 4.	135

Chapter 1
Introduction

1.1 Background

Climbing robot is one of the hottest research topics that gains much attention from researchers. Within this field, most of the climbing robots reported in the literature are designed for climbing manmade structures, such as vertical walls and glass windows [1–9], or structural frames [10–14]. Few climbing robots have been designed to work on natural structures such as trees. Trees and manmade structures are very different in nature. Tree surfaces are seldom flat and smooth, and some trees have soft bark that peels off easily. In addition, the inclined angle on trees is usually not vertical. Hence, most of the climbing methods for manmade structures are not applicable to tree climbing.

To perform tasks on trees, such as tree maintenance, harvesting and surveillance, workers often attach a tool to the end of a long pole to reach the target position. However, this becomes infeasible if the target position is too high. Alternatively, workers can reach the desired position by using an aerial ladder truck. Nevertheless, the use of this equipment is not always possible due to the access limitation such as mountain and rough terrain. In such a case, workers can only climb up the tree to perform tasks. As tree climbing is a dangerous task, robots are expected to assist or replace humans in performing these tasks.

WOODY [15] is one of the climbing robots designed to replace human workers in removing branches on trees. The robot climbs by encircling the entire tree trunk. The size of the robot is thus proportional to the circumference of the trunk. WOODY avoids branches by turning its body and opening the gripper, but the climbed tree trunk should be almost straight. Kawasaki [16] also developed a climbing robot for tree pruning. It uses a gripping mechanism inspired by lumberjacks, and uses a wheel-based driving system for vertical climbing. Same as WOODY, it also needs to encircle the entire tree trunk. Aracil [17] proposed a climbing robot that uses a Gough-Stewart platform to maneuver. It consists of two rings that are joined by six linear actuators through universal and spherical joints at each end. The gripping mechanism also requires the encircling of the entire tree trunk. However, it has greater maneuverability than the aforementioned two robots, and is capable of climbing branchless and curved tree trunks. RiSE V2 [18] is a wall climbing robot that imitates the movement of insects in using six legs to maneuver. This robot has been demonstrated to be capable of climbing trees. As the gripping mechanism only occupies a portion of gripping substrate, the size of the robot is independent of the climbing target. As a result, it is relatively small. However, it does not claim whether it can perform other movement on a tree, such as branch transitions or turning. RiSE V3 [19] is another type of climbing robot designed to climb straight poles with high speed. A hyper redundant snake-like robot [20] can also climb tree. It climbs by wrapping its body around a tree trunk in a helical shape and then rotates its body about its own central axis to roll up the trunk.

For those aforementioned tree-climbing robots, although they work well in their specified purposes, the workspaces are restricted on tree trunks only. Trees with branches and an irregular shape are not considered. They cannot act like arboreal animals such as squirrels to reach any position on irregularly shaped trees with branches. As branches and curvature are presented in many kinds of trees, the application of these robots is strongly restricted.

On the other hand, the autonomous climbing problem on trees is lack of discussions. There is only a single article on the motion planning problem on these tree climbing robots [17]. However, this work merely discussed the local motion planning problem according to local information. Climbing on trees autonomously is challenging as the shape of trees are usually irregular and complex. There are many global motion planning approaches for climbing in manmade structures, such as walls and glass windows [10, 11, 21]. However, these structures are different from trees. The approaches are thus not suitable for tree-climbing problems.

1.2 Motivation

One of the motivations of the work presented in this book is to overcome the workspace limitation of the existing tree-climbing robots. Through billions of years of evolution, many types of arboreal animals have evolved and developed diverse methods to deal with these challenges. The rigorous competition of natural selection process confirms the effectiveness and efficiency of the present solutions in nature. It is believed that the solutions in nature can inspire ideas to solve the captioned technical challenges in certain level. To understand how arboreal animals and artificial robots climb on trees, a comprehensive study and analysis of the tree-climbing methods in both natural and artificial way have been conducted. The study reveals how the wisdom of nature inspires the robot design, and what kinds of designs are solely come from human's intelligent. A novel tree-climbing robot aiming at overcome the workspace limitation is then developed based on the analysis results.

On the other hand, a certain level of autonomous climbing ability of the robot helps reduce the complexity of manipulation required for operation by users. However, the autonomous climbing technology in tree climbing is lack of discussion. In view of that, several approaches in autonomous tree-climbing are proposed in this book, including the sensing methodology, cognition of the environment, path planning and motion planning on both known and unknown environment.

Based on the motivations, the main contributions that can be found in this book are the following:

1. Conducted a comprehensive study and analysis of the methodologies for tree climbing in both artificial and natural way. It provides a valuable reference for robot designers to select appropriate climbing methods in designing tree-climbing robots for specific purposes.
2. Developed a novel extendable continuum maneuver mechanism that allows robots to have high maneuverability and superior extensibility. It opens a new field of applications for continuum mechanisms.
3. Developed a tree surface fastening mechanism with adaptable on a wide range of surface curvature and a wide variety of trees. It permits robots easy to fasten on irregularly shaped trees.
4. Developed a novel tree-climbing robot with distinguished performances. It breaks the workspace limitation of the state-of-the-art tree-climbing robots.
5. Proposed a tree shape reconstruction method based on tactile sensors. It reveals how the realization of an environment can be achieved with limited tactile information.
6. Proposed an autonomous climbing strategy on an unknown shape of trees. It determines the optimal climbing position based on local information and the prediction of a shape of tree in the future path.
7. Proposed a global path planning algorithm on an unstructured 2D manifold (such as tree surfaces). The planned path avoids obstacles and at the same time minimizes the requirement of the fastening force.
8. Proposed a motion planning strategy for continuum type robots to follow a 3D path. The motion planning strategy is generally applicable to any robots with extensible continuum maneuvering mechanism to maneuver in 2D or 3D space.

1.3 Outline of the Book

The book is organized as follows:

Chapter 2 gives a general overview of state-of-the-art tree-climbing robots, to review the cutting edge technologies in this field.

Chapter 3 presents a comprehensive study and analysis on both natural and artificial tree-climbing methods. To provide a valuable reference for robot designers to select appropriate climbing methods in designing tree-climbing robots for specific purposes, the climbing methods are ranked based on the proposed design principles, the climbing methods are evaluated and ranked according to the proposed design principles.

Based on the study and analysis presented in Chapter 3, a novel design of a tree-climbing robot with superior maneuverability and other distinct features is presented in Chapter 4, including its novel mechanical designs, working principles, and control architecture. Numerous climbing experiments are presented to verify the climbing performance of the robot.

In Chapter 5, the optimization of maximizing the fastening force and adaptability of the proposed fastening mechanism are discussed. The relationships among the settings of the mechanism, surface curvature of substrates, and the generated fastening force are studied. The settings for the fastening mechanism are optimized to generate maximal fastening force in a wide range of surface curvature. Numerous experimental results on different kinds of trees are presented to evaluate the actual performance of the mechanism.

Chapter 6 presents the forward and inverse kinematics of the proposed maneuvering mechanism. The workspace analysis is also presented in this Chapter. The workspace represents the admissible fastening positions on a tree surface. The workspace analysis considers both admissible positions and directions of movement on the tree surface.

Chapter 7 proposes a strategy for exploring the unknown environment on trees. Inspired by nature, tactile sensors are adopted to acquire local environmental information. Tree shape approximation, optimal path finding, and motion planning are then developed to make the robot capable of climbing trees automatically. Experimental results are presented to evaluate the performance of the proposed autonomous climbing strategy.

Chapter 8 proposes a global path and motion planning algorithm based on a known environment. An efficient way of formulating the problem is developed and hence the optimal 3D path can be obtained in linear time. A motion planning strategy for 3D path tracking is also proposed. Simulation and experimental results are presented to evaluate the proposed global path and motion planning algorithm.

Chapter 9 concludes the book, and suggests some future research directions.

The book ends with Appendix A, devoted to derivation of equations and the bibliography.

Chapter 2
State-of-the-Art Tree-Climbing Robots

There are several robots developed in the literature that are capable of climbing trees. Some of the robots are specifically designed for tree climbing while some of them are designed for moving in multiple terrains including trees. In this chapter, the details of these robots are introduced to provide a whole picture of the field of the tree-climbing technology.

2.1 WOODY

The WOODY project [25] have been started since 2004 and three generations of the prototype have been developed. They were developed in Sugano Lab at Waseda University, Japan. The main motivation of the development of WOODY is forest preservation. In forests, too many branches presented on trees block the sunlight and kill the glass and frutex on the ground. The beauty of the forest will be affected. In addition, if trees present too many branches, heavy rain and snow will make trees overload and breakdown. The trees will then fall down and kill forestry workers. As a result, pruning should be conducted frequently. WOODY is then designed to assist forestry workers to remove branches from trees.

WOODY is a manually controlled robot. It fastens onto a tree by encircling an entire tree trunk with its arm. It climbs vertically by extending and contracting its body using threaded rod mechanism and at the same time releases and encloses its upper and lower arms alternatively. As WOODY's arms have to encircle the entire tree trunk, the size of the robot is proportional to the circumference of the trunk. There is a cutter installed on the top of the robot body for branch cutting. Active wheels rolling in horizontal are installed on the interior of each arm, allowing WOODY turns about a tree trunk to go to proper position for pruning.

The mechanical structure allows WOODY to climb on a straight tree trunk only. As the trunk of the target climbed trees, i.e., cedar and cypress, are almost straight and the branches will be removed by the equipped cutter before the robot passing through, the design of WOODY is sufficient to perform the task.

2.2 Kawasaki's Pruning Robot

Kawasaki [16] developed a tree-climbing robot for pruning and it is conducted at Gifu University, Japan. The first prototype is published in 2008. Similar to the purpose of WOODY, the task is pruning on a straight tree trunk.

In their concerns, the main motivation of pruning is to keep the lumber of the pruned tree has beautiful surface without gnarl and homogeneous quality with well-formed annual growth ring so that the trees can produce worth wood.

Although the Kawasaki's pruning robot and WOODY share the same mission, the fastening and maneuvering mechanisms are different. Kawasaki aims to make a lightweight and fast climbing pruning robot. The fastening mechanism of the pruning robot is inspired by the approach of lumberjacks that uses self-weight locate the center of gravity exterior against tree to produce pressing force on the upper and lower fulcrums. The robot can then hold on the tree without pressing mechanism against a tree. The fulcrums of the robot are form by four active wheels that set at regular intervals around the tree, which one pair for upper side and the other for lower side. The active wheel contributes the movement of the robot that the wheel-driven mechanism can make the robot climb fast. Each active wheel is driven by motor independently through a worm gear to avoid backdrive. In the first prototype of the pruning robot [16], wheels are aligned vertically and hence the robot can only perform moving up and down motion. In the second prototype [24], the wheels are capable of steering actively. As a result, except the linear movement, the robot can switch to spiral climb by adjusting the orientation of wheels. The climbing motion can be adjusted according to the situation in real-time so as to improve the moving efficiency.

2.3 Seirei Industry's Pruning Robot

AB232R [26] is a commercialized pruning robot promoted by Seirei Industry Co. LTD. It is a wheel-driven robot with fixed orientation of wheels and hence it can perform a constant spiral climb with constant speed only. A branch cutter is installed on the top of the robot. The robot utilizes the constant spiral climbing up motion to make the cutter pass through all the surface of a tree trunk for pruning all the branches. The robot can simply be commanded to move spirally upward and downward only. As the robot has to encircle an entire tree trunk and the pressing force to hold its body on the tree is generated by a passive preloaded mechanical spring, the diameter of the target trees is limited in a certain range. The range is proportional to the size of the robot. Hence, Seirei Industry provides two models target on different range of trunk diameter, i.e., AB232R for diameter 70-230mm and AB351R diameter 150-350mm respectively.

2.4 TREPA

Reinoso, at the Miguel Hernandez University, proposes using Gough-Stewart platform with 6 DOF as a climbing platform called TREPA. The concept is first proposed in 1999 [27]. They claimed that the platform can be applied for climbing metallic structures and cylindrical structures like palm tree by different configurations. The prototype was then built in 2003 [28] and demonstrated the climbing motion along trunks of palm trees.

The objectives of making a robot for palm tree climbing are branch trimming and fumigation as they discovered that most palm trees on the Spanish Mediterranean coastlines are affected by disease and there is not enough expert operators to do this dangerous task.

TREPA consists of two hexagonal rings that are linked with six linear actuators through universal and spherical joints at each end respectively. The parallel driving mechanism allowing it has great load capacity and the six DOFs permit the robot to move to a certain position and orientation in the workspace. Its high maneuverability allows TREPA not only capable of climbing a straight trunk, but also capable of climbing a palm tree with a certain range of bending. The gripping mechanism requires the robot to encircle an entire tree trunk. Each hexagonal ring have three pressing system in linear motion toward the center of tree trunk to fix the robot on a tree.

On top of that, they also propose an autonomous climbing algorithm that permits TREPA to climb along an unknown shape of a palm tree. It is achieved by detecting the distance between the hexagonal ring and trunks by three regularly separated ultrasonic sensors installed on the hexagonal ring.

2.5 RiSE

The goal of the RiSE project is to create a bio-inspired climbing robot to walk on land and climb on vertical surfaces that has potential to apply in search and rescue, reconnaissance, surveillance or inspection applications. This project is funded by the DARPA Biodynotics Program. The involved parties include the Boston Dynamics Inc., Lewis and Clark University, University of Pennsylvania, Carnegie Mellon University, Stanford University, and U.C. Berkeley. RiSE V1 [22], RiSE V2 [18], and RiSE V3 [19] are part of the overall RiSE Project, that designed and built by Boston Dynamics. They are capable of climbing on tree trunks or wooden straight poles vertically.

RiSE V1 was announced in 2005, and RiSE V2 is the improved version of RiSE V1 which shares similar architecture. It imitates the movement and climbing method of an insect, using six legs to maneuver and spines for climbing. Each leg has two active joints actuated by two motors independently and hence there is total of twelve actuated degrees of freedom. Spine is installed at the end of each leg and a compliance structure is applied to help the robot attach tightly on a climbing surface.

Through a tripod pattern of the leg motion, the robot can climb on trees vertically. As the gripping mechanism only occupies a portion of a climbing substrate, the size of the robot is independent to the size of climbing target. As a result, it can be relatively small.

RiSE V3 [19] is another type of climbing robot that uses quadrupedal mechanism for climbing. It is designed to climb wooden straight poles with high speed. The design and motion of the legs are similar to that of RiSE V2. However, it needs to grip half of a gripping surface to ensure stable climbing and hence the size of the robot is proportional to the size of climbing target.

2.6 Modular Snake Robot

A hyper-redundant snake-like robot can imitate living snakes to move in different ways such as rolling, wiggling, and side-winding so as to move around different terrains. A modular snake robot [20] developed at Carnegie Mellon University, named Uncle Sam, can even climb on a straight tree trunk vertically from a ground. The climbing motion is totally different to that of the tree climbing method used by living snakes. The snake robot wraps itself around a tree trunk in a helical shape and then rotates its body about its own central axis to roll up the trunk. By using this climbing method, the length of the robot must be long enough to encircle the whole tree trunk. Although it was demonstrated the capability of climbing a straight tree trunk only, the mechanism has potential to perform branch to branch transition as long as the length of the robot is long enough to wraps between two branches.

Chapter 3
Methodology of Tree Climbing

3.1 Tree-Climbing Methods in Nature

Arboreal habitats are complex environments that pose numerous challenges to arboreal animals. These include climbing on cylindrical branches with variable diameters and inclined angles, moving in narrow spaces and facing obstacles. Through billions of years of evolution, many types of arboreal animals have evolved and developed diverse methods for moving around in complex arboreal environments.

Many tiny animals, such as snails and worms, adopt wet adhesion to fasten themselves on trees. The wet adhesion includes the capillary adhesion and suction mechanisms. This method only provides limited adhesion force, but it is sufficient to support such tiny creatures. Although these animals use this fastening method, their locomotion is different.

Snails have a single foot and move by pedal wave locomotion (waves of muscular contraction) [64]. The area of adhesion (the entire foot) is relatively large, which provides a relatively large adhesion force to support their relatively heavy bodies. However, this method allows slow movement only.

Caterpillars have multiple feet that act as suctions to adhere on substrates. They move by using a sinusoidal gait which is quite similar to the pedal wave locomotion of snails.

Another common fastening method observed in tiny animals is Van der Waals force [65], which is commonly adopted by insects. Insect's feet have pulvilli (or pad) that can adhere to a smooth surface by Van der Waals force (dry adhesion). Some insects also have micro claws on their feet to hook themselves onto non-smooth substrates.

As for the locomotion, insects use a tripod gait, whereby at least three feet are in contact with the substrate at any time to make adhesion more stable.

Inchworms (also called loopers) use both wet adhesion (on the hind foot) and micro claws (on the fore foot) to attach themselves to surfaces. Their locomotion is unique. When they move, they use either the fore foot or the hind foot to attach to a substrate, and then bend their body to move the other foot to a new position alternatively. This can be treated as a kind of bipedal locomotion. Although inchworm locomotion is fast, the bending of the body makes the center of mass away from the climbing substrate, which generates a large pitching force. As inchworms are lightweight, the pitching force doesn't affect much of the movement.

The various climbing methods used by tiny animals treat branches as a flat surface rather than a cylindrical shape, and thus nearly any size of branch can be climbed.

Small size animals such as squirrels and birds use claw penetration method to fasten on a tree. Claws can be used to interact with rough substrates. To generate a large gripping force, the angle of grip should be as large as possible. The claw-gripping method can be used to fasten onto smooth or rough branches with a gripping angle less than 180 degrees as illustrated in Fig. 3.1(b), which allows squirrels to climb large tree trunks.

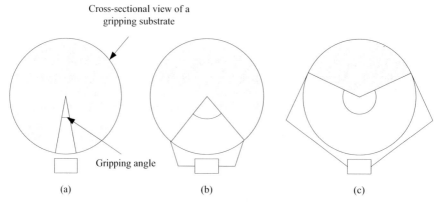

Fig. 3.1 Illustration of the gripping angles by using different methods. (a) Wet adhesion or Van der Waals force; (b) Claw gripping; (c) Interlock method.

As quadrupedal animals, squirrels use their limbs for maneuvering, usually through pulse climbing, which is a type of dynamically stable locomotion that allows very fast movement.

Birds usually do not maneuver continuously throughout a tree, but instead move by hopping along a branch or flying.

Large size animals such as primates do not have sharp claws, and instead use interlock method to fasten on trees. They hold their bodies on a tree by encircling more than half of a branch with the forelimbs as illustrated in Fig. 3.1(c) to pull the body toward the branches while both hind limbs push the body upward. This interlock method provides a large gripping force and hence is more suitable for large animals [60]. This fastening method depends on the angle of the frictional force, and thus depends on the diameter of the branch. As a result, climbing on a larger branch results in a reduced gripping force.

Primates maneuver by moving diagonally opposing limbs at the same time, resulting in symmetrical footfall patterns [62].

Snakes also adopt frictional gripping for tree climbing, but the gripping mechanism is totally different to that of primates [59, 61]. As snakes are long and thin, they fasten on a tree by sticking their bodies into the deep vertical furrows in the bark and then move by internal concertina locomotion [66]. The frictional area (contact area) with this method is large and the center of mass is extremely close to the substrate, which generates sufficient fastening force. However, it means that snakes can climb very rough-barked but not smooth-barked trees [63].

There are numerous approaches to tree climbing in the natural world, each of which is suitable for particular animals in a certain kind of arboreal habitat. Table 3.1 summarizes these natural tree-climbing methods.

3.2 Artificial Tree-Climbing Methods

Arboreal animals provide many ideas for the design of tree-climbing robots. There are several tree-climbing robots have been developed as mentioned in Chapter 1. Most of these designs have adapted climbing approaches from natural world with certain modifications, simplifications and creations, instead of implement directly to suit specific purposes and artificial design spaces.

WOODY, TREPA and Kawasaki's pruning robots apply frictional gripping to attach to a tree while RiSE V2 and V3 use claw penetration method. The locomotion of RiSE V2 imitates insects' locomotion (tripod gait) while the locomotion of RiSE V3 imitates primates' locomotion (diagonal footfall patterns). WOODY and TREPA move by extending and contracting their bodies in a manner similar to inchworms locomotion, except that the center of mass is closer to the gripping surface. TREPA's maneuvering method even allows the body to have certain bending to change the maneuvering direction which is similar to the motion of inchworms.

Scientists have also created climbing robots from completely new climbing method. For example, the Kawasaki's pruning robot uses wheel-driven method to move, which would never be seen in nature. This wheel-driven method allows very fast movement. The locomotion that the snake-like robot uses to climb a pole is also creative that real snakes never move like this. The robot encircles the pole by winding its body in a helical shape to fasten itself on the pole using frictional gripping. It climbs by rotating its body along its own central axis, which can be treated as a kind of wheel-driven locomotion. This locomotion is faster than that of real snakes.

Table 3.2 summarizes the artificial tree-climbing methods.

Table 3.1 Summary of the tree-climbing methods applied in nature.

Fastening method	Maneuvering method	Examples	Advantages	Disadvantages
Wet adhesion	Pedal wave	Snail	Large adhesive force	Slow locomotion
	Sinusoidal gait	Caterpillar		
	Body bending	Inchworm	Fast locomotion	Suffer from pitching force
Van der Waals force	Tripod gait	Insect	Stable locomotion	Provide less fastening force
Claw penetration gripping	Hopping, flying	Bird	Wide range of gripping curvature; Fast locomotion	Fastening force depends on the property of gripping substrate
	Pulse climbing	Squirrel		
Frictional gripping	Diagonal footfall patterns	Primate	Provide large fastening force	Need to encircle more than half of a gripping surface
	Internal concertina locomotion	Snake	Locomotion adaptive to many kinds of terrain	Restricted on bark with deep vertical furrows; Slow in motion

3.2 Artificial Tree-Climbing Methods

Table 3.2 Summary of the artificial tree-climbing methods.

Fastening method	Maneuvering method	Examples	Advantages	Disadvantages
Claw penetration	Tripod gait	RiSE V2	Stable gripping, wide range of gripping curvature	Slow locomotion
	Diagonal footfall patterns (Quadrupedal)	RiSE V3	Fast locomotion	Not stable
Frictional gripping, En-circle	Wheel-driven	Kawasaki's pruning robot	Fast locomotion	Low maneuverability
	Extend-contract	WOODY	Stable grip	Gripping range is limited
	Extend-contract with bend	TREPA	High maneuverability	Complex
	Rolling in helical shape	Snake-like robot	adaptive to different terrains	Complex, multiple actuators

3.3 Design Principles for Tree-Climbing Robots

The previous sections have introduced natural and artificial tree-climbing methods. In this section, design principles for making tree-climbing robots are proposed to evaluate the climbing methods. It also help robot designers select appropriate fastening and maneuvering methods in designing tree-climbing robots for their specific purposes. To design a tree-climbing robot, six major design principles should be considered: *high maneuverability*, *high robustness*, *low complexity*, *high adaptiveness*, *small size*, and *high speed*.

High Maneuverability: As the structure of trees are irregular and complex in terms of geometry, reaching a wide range of positions requires high maneuverability. Simplified artificial maneuvering methods always provide limited degrees of freedom and hence restrict the climbing space such as capable of climbing around tree trunk only. A maneuvering method with higher maneuverability must be devised if the robot is to be applied to a wide range of climbing workspaces such as capable of climbing to the crown of a tree. However, mechanical designs with a higher maneuverability always result in higher complexity.

High Robustness: Robustness represents the ability to hold on a tree without falling off. High robustness is preferred in general, especially when the robot itself is heavy or equipped with heavy payload. It is related to the fastening methods and maneuvering methods, as some locomotion will generate extra dynamic force for which the fastening method cannot compensate. For example, making the center of mass of a robot higher than the branch on which it is to climb will increase the tendency of pitching backwards or toppling sideways [59, 61].

Low Complexity: Mechanical and control complexity are important considerations. If a robot is mechanically complex, it might increase the cost, size and the weight of the robot and the susceptibility to fail. Complex in control also reduces the robustness of a robot as failure is more likely to occur. In addition, a powerful processor should be deployed to handle the complex control algorithm which has potential to increase the cost and energy consumption. As a result, low complexity is preferred in general.

High Adaptability: Arboreal habitats are varied. There are different kinds of trees with different surface textures. In addition, the shape of the surface of trees is irregular and the curvature is varied in different positions. As a result, a fastening method should thus be selected that can adapt to a range of expected workspaces. Higher adaptability defines as the larger range of the environment that is capable of fastening on. To adapt a larger range of environment, the fastening mechanism and the control usually become more complex.

Small Size: A small size provides many advantages, such as the ability to climb in narrow spaces and probably lightweight which induces fewer dynamic force and

3.4 Ranking of the Tree-Climbing Methods

preload. In addition, small size robot is easy to transport. It is important as the transportation of heavy and large machinery is not possible in some environments in which it might be used such as mountain.

High Speed: Fast climbing is preferred in many applications as it can shorten the time of operation and hence improve the productivity. Nonetheless, a faster speed may also reduce the robustness of a robot.

It can be seen that these design principles are correlated at certain levels. The potential relationship among the design principles is summarized in (3.1). The up and down arrows represent the increase and decrease of the items respectively. Trade-offs will need to be made among them base on a specific application of the robot.

$$\left.\begin{array}{c}Maneuverity \uparrow \\ Adaptability \uparrow\end{array}\right\} \Rightarrow Complexity \uparrow \Rightarrow \left\{\begin{array}{c}Size \uparrow \\ Robustness \downarrow \Leftarrow Speed \uparrow\end{array}\right. \quad (3.1)$$

3.4 Ranking of the Tree-Climbing Methods

Table 3.3 ranks the natural and artificial climbing methods reviewed in the previous sections based on the six design principles. This will help with the selection of the most appropriate climbing method in designing a tree-climbing robot. The rank has six levels labeled from 1 to 6 representing the performance from the worst to the best. Higher rank refers to higher adaptability, robustness, maneuverability, smaller size, and lower complexity. The details of the ranking are explained as follows.

3.4.1 Maneuverability

For the existing extend-contract and wheel-driven methods, they can maneuver in one dimension only, so the ranking is the lowest. The body bending method can achieve arbitrary position in two-dimensional manifold in each motion, so the ranking is higher than the above method. The extend-contract with bend method can even extend the motion in to three-dimentional space in each motion, hence the ranking is much higher. As for the tripod and quadruped gait method, they can achieve three-dimentional space motion with certain redundancy. The redundancy results in more than one solution to achieve a same goal which provides more feasibility for obstacle avoidance. Hence the ranking is higher than the extend-contract with bend method. The wave form and rolling in helical shape methods adopt hyper redundant actuators and hence they have a much higher redundancy in the motion. As a result, they are ranked as highest.

Table 3.3 Ranking of the fastening and maneuvering methods in tree climbing (1=worst, 6=best).

Design principles	Adaptability	Robustness	Size	Complexity	Maneuverability	Speed
Fastening method						
Wet adhesion	6	2	6	2		
Van der Waals force	6	1	6	1		
Claw penetration	4	5	4	4		
Frictional gripping	1	6	1	6		
Maneuvering method						
Wave form		6	1	1	6	1
Body bending		1	5	4	2	4
Extend-contract		3	6	5	1	4
Extend-contract with bend		3	5	4	4	4
Tripod gait		5	2	2	5	2
Quadruped gait		4	3	3	5	4
Wheel-driven		6	6	6	1	6
Rolling in helical shape		6	1	1	6	6

3.4.2 Robustness

In view of the fastening method, robustness means the magnitude of the fastening force. The larger the fastening force, the greater the robustness can be achieved. The frictional gripping interlocks with a branch and hence can provide infinite force. It will be pulled out only if the branch is broken by the pulling force. The claw penetration method actually interlocks with the tree bark to a certain depth, which will break more easily than a whole branch. Hence, the robustness of claw penetration is ranked lower than frictional gripping. As for the wet and dry adhesion methods, they only adhere to the surface of a tree. It will be highly affected by the dustiness of the surface. Hence the fastening forces using these methods are ranked lower than frictional gripping and claw penetration. According to [71], the range of Van der Waals force (dry adhesion) and capillary force (wet adhesion) is about 10^{-11} to 10^{-9} and 10^{-7} to 10^{-3} kgf respectively. In experiments, Jiao [73] found that the adhesion force of an insect leg (Van der Waals force) is $1.7 - 2.2 mNmm^{-2}$. Kim [72] found that for a snail, the total maximum capillary force is $3.1 mNmm^{-2}$. As a result, the wet adhesion force is larger than Van der Waals force.

As for the maneuvering method, robustness means the magnitude of the dynamic force generated by the maneuvering motion. The smaller the dynamic force, the greater the robustness can be achieved. Wave form, rolling in helical shape, and wheel-driven methods have the highest rank as these motions do not change the center of mass relative to the fastening position, which is most stable. A tripod gait has a higher rank than quadrupedal gait because there are three supporting points when maneuvering in tripod gait, while there are only two supporting points in a quadrupedal gait. As for the extend-contract, and extend-contract with bend methods, although the contraction motion makes the center of mass close to a gripping substrate which eliminates the pitch back moment, the extension motion moves the center of mass apart from the fastening position, which increases the side-toppling force and pitch-back moment. As a result, they are ranked lower than the quadrupedal gait method. Regarding the bending motion, the action of straightening the body raises the same issue as the extension motion in the extend-contract method. The body bending makes the center of mass out of the gripping substrate, and it will increase the pitch-back moment and side-toppling force. As a result, it has the lowest rank.

3.4.3 Complexity

The complexity represents both the fabrication complexity and the control complexity of the mechanism. For the fastening method, the control of frictional gripping is the simplest as it uses the interlock method. Putting robot limbs in the proper position can achieve the fastening effect. Although the claw penetration method can be fabricated easily, it is complex to control as the claw should be positioned directionally to the substrate to generate optimal gripping force [35]. Hence the rank is lower than the frictional method. The wet adhesion method requires special liquid between the object and the substrate [68]. The continuous provision of this consumable substance makes the application difficult to apply. In addition, to use a suction force, it is necessary to remove the air between the pad and substrate to reduce internal pressure, which is not easy on irregular tree surface. As a result, the rank is lower than the claw penetration method. As for the Van der Waals force, the micro/nano meter scale biomimetic pad is difficult to make. Although there are a lot of studies focusing on it, such as [1] [69] [67], the technology is still not mature enough. The adhesive force will degrade with time due to dust and dirt [70], so its rank is the lowest.

For the maneuvering method, the complexity of maneuvering is defined as the control effort. The wheel-driven method maneuvers by a continuous rolling motion. One control command is enough for maneuvering, so the rank is highest. The extend-contract method consists of two motions, i.e., extend and contract. A two-motion step is needed, which is somewhat more complex than the wheel-driven method and hence the rank is lower. The body bending and extend-contract with bend methods need to control the bending in three-dimensional space. It needs six control inputs to define the position and orientation in three-dimensional space which is more complex to control. In a quadrupedal gait, it needs to control four limbs. In addition, the motion of the limbs must be synchronized when moving. It is much more complex than the above method. In tripod gait, the situation is similar to the quadrupedal gait, but it needs to control six limbs and hence the rank is lower than the quadrupedal gait method. As for the wave form generation and rolling in helical shape, they are achieved by using hyper redundant actuators. Hence they rank the lowest.

3.4.4 Adaptability

The adaptability of the fastening method represents the range of admissible gripping substrate and large range is preferred. To compare the adaptability fairly, the fastening mechanism of different fastening method is assumed to be in the same size. Frictional gripping method requires gripping over half of the surface (gripping angle larger than 180 degrees). The range of claw penetration method covers all the range of frictional gripping method and even capable of fasten on a surface with gripping angle smaller than 180 degrees. As a result, the claw penetration method is ranked higher than the frictional gripping method. The wet adhesion and Van der Waals force can adhere on any shape of substrate such as concave surface that the claw penetration method cannot achieve. Hence the ranking of wet adhesion and Van der Waals force are highest.

3.4.5 Size

As for the fastening method, the size of a robot using frictional gripping is similar to the diameter of the gripping substrate, while by using the claw penetration, under half of the size is adequate. As for wet and dry adhesion, they are independent of the size of the gripping substrate. As a result, the ranking of size for the fastening method is same as the adaptability for the fastening method.

The size of the maneuvering methods is related to the complexity of the mechanism. It can be quantified by the number of actuators required. The simpler the mechanical structure, the smaller the robot can be. A wheel-driven method is the simplest mechanism, as it maneuvers by rolling. One control command is enough, so the rank is highest. The extend-contract method is a one DOF motion, which can be achieved by an actuator. As a result, the rank is the same as the wheel-driven method. The body bending and extend-contract with bend methods needs to control the bending in three-dimensional space. It needs at least six DOF to define the position and orientation in three-dimensional space. In a quadrupedal gait, each limb only has two DOF (which can be more in actual cases), so eight actuators are required in total. The concept is the same for a tripod gait, twelve actuators are required. As for wave form generation and rolling in helical shape, it is achieved by hyper redundant actuators. Hence their ranks are the lowest.

3.4.6 Speed

The speed of the maneuvering method is determined by how far a robot can move forward in one climbing gait. Since the wheel-driven and rolling in helical shape methods do not have a concept of climbing gait and can move continuously, their ranks are the highest. The distance moved in a climbing gait is equal to the wave length of the motion. The wave length is usually two times of the distance between the adjacent legs d as illustrated in Fig. 3.2. To make a fair comparison, the distance is compared by dividing the maximum length of the robot L, that is,

$$\frac{2d}{L} \qquad (3.2)$$

In each of the climbing gait, body bending, extend-contract, and extend-contract with bend, and quadruped methods can move same as the maximum length of the robot and their ranks are highest. As for the tripod gait, it can only move forward two third of its maximum length. As a result, it is ranked lower than the above method. As for the wave form method, since it moves by multiple legs, it ranks the lowest according to (3.2).

In summary, the performance of the existing tree-climbing robots based on the design principles can be evaluated by radar charts as illustrated in Fig. 3.3. In the

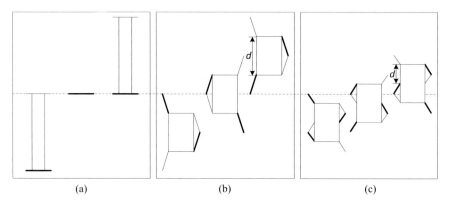

Fig. 3.2 Climbing gaits of different maneuvering methods. (a) Extend-contract gait; (b) Quadrupedal gait; (c) Tripod gait.

figures, blue and red parts represent the performance of the fastening and maneuvering method respectively, while the black lines represent the integrated performance. It clearly shows the strengths and weaknesses of the robot designs. The radar chart can also be deployed to design and evaluate a new robot by different combination of the fastening and maneuvering methods.

3.5 Summary

A comprehensive study of the tree-climbing methods in both natural and artificial aspects has been undertaken. The major fastening and maneuvering methods for an arboreal environment have been introduced and discussed. The six major design principles were then proposed for designing a tree-climbing robot. The performances of the natural and artificial climbing principles have been ranked based on the design principles. It is a good reference to help with the selection of the most appropriate climbing principles in designing a tree-climbing robot for specific purposes.

3.5 Summary

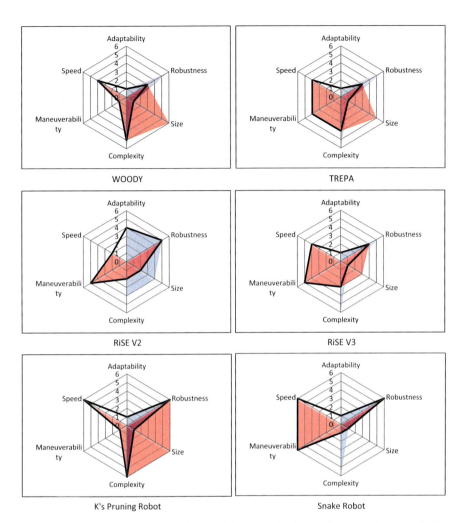

Fig. 3.3 Performance evaluation of the existing tree-climbing robots based on the design principles.

Chapter 4
A Novel Tree-Climbing Robot: Treebot[1]

4.1 Objectives

Based on the limitation of the existing tree-climbing robots, one of the motivations of the work presented in this book is to develop a novel type of tree-climbing robot that can assist or replace humans in performing tasks on trees. The robot should be applicable to various pursuits such as harvesting, tree maintenance and observation of arboreal animals. To achieve this goal, the robot should meet several requirements:

1. High maneuverability. The robot should be capable of turning, transiting, and climbing irregularly shaped trees to enhance the climbing workspace.
2. Adhesion to tree surfaces with a wide range of curvatures. It enables the robot to climb from a large tree trunk to a small branch without needing to replace the gripper.
3. High payload capacity. It allows the robot to carry necessary equipment for performing tasks.
4. Lightweight and compact. Making a robot portable and easy transport is important because the transportation of heavy machinery is not possible in some environments in which it might be used such as mountain.
5. Energy saving. As a self-contained field robot, power is provided by an integrated battery. As the capacity of battery is limited, an energy saving design is crucial to prolong the robot's working hours.
6. An autonomous climbing ability. A certain level of autonomous climbing ability of the robot helps reduce the complexity of manipulation required for operation by users.

---

[1] Portions reprinted, with permission from Tin Lun Lam, and Yangsheng Xu, "A Flexible Tree Climbing Robot: Treebot - Design and Implementation", Proceedings of the IEEE International Conference on Robotics and Automation. ©[2011] IEEE.

4.2 Approach to the Robot Design

The design principles needed to be considered with priority included maneuverability, adaptability, robustness, and size. Table 3.3 can be used as a reference to design the tree-climbing robot.

In terms of fastening method, only claw penetration method has high ranks in all considered design principles. As a result, this method is adopted in the robot design. As for the maneuvering method, the table shows that the extend-contract with bend methods have satisfied performance in view of the considered design principles. As a result, the extend-contract with bend method is adopted. According to Table 3.3, the expected performance of a tree-climbing robot with the use of claw penetration method for fastening and extend-contract with bend method for maneuvering is shown in Fig. 4.1. Although the maneuverability of the extend-contract with bend method is not the highest, it is enough to fulfill the maneuver requirements. In order to further improve the performance, a novel maneuvering mechanism is proposed. It is still using the extend-contract with bend motion but fewer actuators are required. Hence the size and complexity can further be reduced.

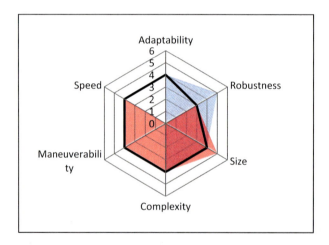

Fig. 4.1 Expected performance of a tree-climbing robot with the use of claw penetration method for fastening and extend-contract with bend method for maneuvering.

4.3 Structure

Fig. 4.2 shows the overall structure of the proposed tree-climbing robot "Treebot". Treebot is composed of three main elements: a *tree gripper*, a *continuum body* and a *semi-passive joint*. Two grippers connect to the ends of the continuum body respectively, and the semi-passive joint is installed between the body and the front gripper.

4.3 Structure

Treebot has three active DOFs, i.e., the continuum body and the two passive DOFs, i.e., in the semi-passive joint, which is a kind of underactuated robot. On the other hand, only five actuators are used, two for the motion of the grippers and three for the motion of the continuum body. Several *sensors* are also installed for exploring the environment.

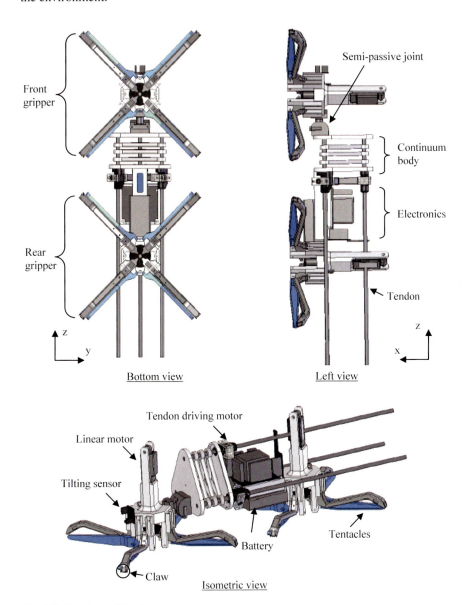

Fig. 4.2 Structure of Treebot.

4.3.1 Tree Gripper

The proposed gripper is designed to fasten onto a wide variety of trees with a wide range of curvatures. There are many innovative approaches for generating adhesive force, such as vacuum suction [4, 8, 51, 52], magnetic attraction [5, 6, 31, 49, 50], elastomeric adhesive [7, 53], electroadhesion [2] and fibrillar adhesion [1, 3]. These methods work well on manmade structures such as vertical walls and glass windows that are smooth and flat. However, they are not applicable to tree surfaces, which are completely different in nature. Observation of the arboreal animals indicates that the claw gripping method is reliable on tree surfaces, and this method is thus adopted in the proposed tree gripper to generate the fastening force.

The gripper is composed of four claws equally separated by 90 degrees that permits omni-directional gripping about its principal axis. Fig. 4.3 shows the mechanical design of the gripper in different views. The design is somehow like birds' feet. The gripper should be appressed to the tree surface (the center of the gripper makes contact with the gripping surface and the principal axis of the gripper is collinear with the surface normal) to generate maximal fastening force. Each claw is composed of two parts, Phalanx 1 and 2, and has surgical suture needles (spine) installed at the tips. The claws adopt a two-bar linkages mechanism to generate the optimal direction of acting force. Each gripper has one linear motor that actuates all claws. A pushing plate is mounted at the end of the linear motor. The gripping mechanism is illustrated in Fig. 4.4. When the linear motor extends, the plate pushes all the Phalanx 1 downward and hence makes the Phalanx 2 upward. The spring on Joint A is compressed. This motion pulls spines off the gripping substrate. When the linear motor contracts, the compressed springs on Joint A generate a force to push the claws back into the gripping substrate. As the gripping force is generated by the compressed springs only, static gripping can be achieved with zero energy consumption. A constant force spring (a flat spiral spring) is adopted to ensure that the force is independent of the claw traveling angle. In addition, as the mechanism of each claw is independent, the claws are capable of traveling in different angles. This ensures that all of the claws penetrate into the gripping substrate to generate the maximal force even if it has an irregular shape.

Tentacles are installed beside each claw. These have various functions, including acting as tactile sensors, helping to ensure that the gripper is appressed to the tree surface, and helping the claws to retract from the gripping substrate. The gripper allows omni-directional gripping about its principal axis so that no additional orientation actuator or control is needed. Additionally, the gripper is actuated by one motor only, which makes it light, compact, and easy to control.

Fig. 4.5 shows the prototype of the proposed tree gripper. The gripper is 130mm (height) × 160mm (width) × 160mm (length) in size and weighs 120 grams. The torque on each spiral spring at Joint A is around 50Nmm and the preload torque on each spring at Joint B is around 35Nmm. The linear motor can open the gripper by around 20N pushing force. It can be observed from the figure that the claws can travel in different angles to adapt the curved surfaces.

4.3 Structure

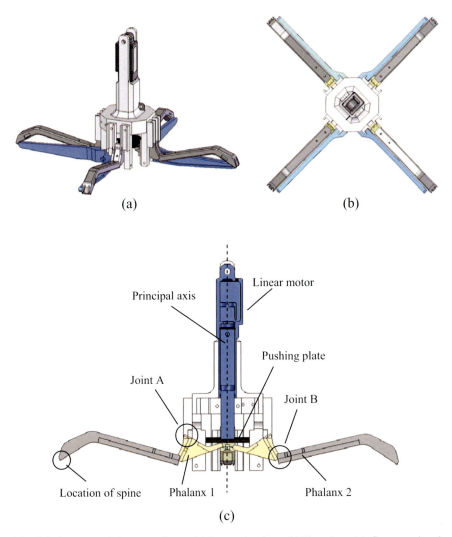

Fig. 4.3 Structure of the tree gripper. (a) Isometric view; (b) Top view; (c) Cross-sectional view.

4.3.2 Continuum Body

There are many types of continuum manipulators that utilize pneumatic-driving [36–38] or wire-driving mechanism [40, 41]. Most of them are capable of bending in any direction and some of the pneumatic-driving manipulators are even capable of extending to a certain extent. Most of the researches use the continuum structures as robot arms, but seldom researchers have realized that it can also be applied to maneuver. The continuum mechanism is a compliant structure, as it does not

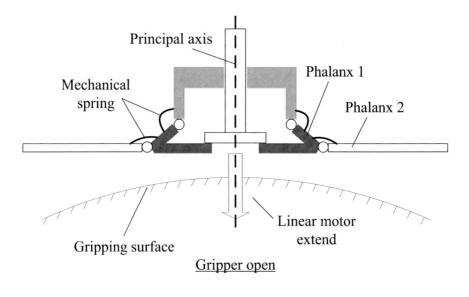
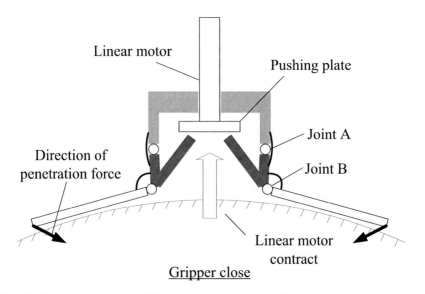

Fig. 4.4 Gripping mechanism of the tree gripper in cross-sectional view.

4.3 Structure

Fig. 4.5 Prototype of the tree gripper.

contain fixed joints [39]. Its inherent passive compliance is of particular benefit for maneuvering in an irregular arboreal environment, as it often eliminates the need for complex force sensing and feedback control [34]. For climbing purposes, the manipulator should be compact and lightweight. There are many types of continuum manipulator, but none of them fulfills all of these requirements. The existing continuum manipulators need to connect to a large external device that contains wires, drivers, motors, or air pumps. Although some wire-driven continuum manipulators [40, 41] have the potential to be more compact and lightweight, they are not extendable. Extendibility is important, as Walker [42] shows that the inclusion of extension ability for continuum manipulators highly extends the workspace.

Due to the limitations, a novel design of a continuum manipulator with both bendable and extendable abilities is proposed as the robot's body for maneuvering. The continuum body is a type of single section continuum manipulator [30] with a novel mechanism. It has high degrees of freedom (DOFs) and a superior ability to extend that the existing manipulators cannot achieve. The continuum body can extend more than ten times its contracted length. By comparison, OCTARM V [37], has only 75 percent extension capability. This allows the robot to maneuver in complex arboreal environment. The continuum body has three DOFs that can extend and bend in any direction. In Treebot, it acts as a maneuvering mechanism to place one end of the gripper on a target position, allowing the robot to reach many places on a tree. The locomotion of Treebot is similar to that of an inchworm robot [31], except that the moving motion is achieved by body extension and contraction, rather than body bending. As the extension and contraction maneuvering mechanisms place the center of mass of Treebot close to the climbing surface, a smaller pitch-back moment is produced when climbing. Jiang [32] also proposed another type of inchworm-like robot for planar maneuvering. Similar to our approach, it moves by extending and contracting its body but no bending motion is allowed. Lim [33] proposed a pneumatic-driven extension and contraction moving mechanism that is able to be bent passively. However, it is only suitable for inner pipe maneuvering.

The inherent passive compliance of the continuum body allows it to be sheared in 2-DOF along x- and y-axes and be twisted about z-axis by external force. The definition of the coordinate system is shown in Fig. 4.2. The compliance is useful for adapting to the irregular shape of trees, as it eliminates the need for complex force sensing and feedback control [34]. The proposed continuum manipulator is a self-contained module with integrated actuators, and thus no external control box is required. This makes the continuum body compact and lightweight.

Fig. 4.6 shows a computer-aided design (CAD) model of the proposed continuum body. It is composed of three mechanical springs that are connected in parallel. The distances between the center of the continuum manipulator and the springs are equal, and the springs are equally separated by 120 degrees. One end of each spring is fixed on a plate, and the other end has no fixed connection. The springs pass through a plate that contains three DC motors to control the length of the springs between the two plates independently. By controlling the length of each spring, the continuum manipulator can perform bending and extension motions. Commonly, the number of actuators required for each section of the continuum manipulator is more than the number of admissible degrees of freedom. However, in the proposed structure, only three actuators are used but three DOFs are provided. This structure provides the maximum DOF with the minimum number of actuators. Fig. 4.7 shows the prototype and illustrates some of the admissible motions of the continuum body. The actuation mechanism is similar to a rack and pinion mechanism, which allows the unlimited extension of the continuum manipulator theoretically. In practice, extension is limited by the length of the springs. Each spring can be treated as a bendable rack, and is only allowed to bend in any direction but not to compress or extend to keep a constant gap distance within which the pinion can drive. Maintaining the springs at constant intervals across the body is important to retain a uniform shape, and several passive spacers are installed in the middle of the body for this purpose. The maximum distances between the spacers are constrained by wires.

The inherent passive compliance allows the continuum body to be sheared along the x- and y-axes and twisted about z-axis by external force, as shown in Fig. 4.8. This passive compliance results from the bendable characteristics of the mechanical springs. The amount of compliance increases when the continuum manipulator extends.

4.3 Structure

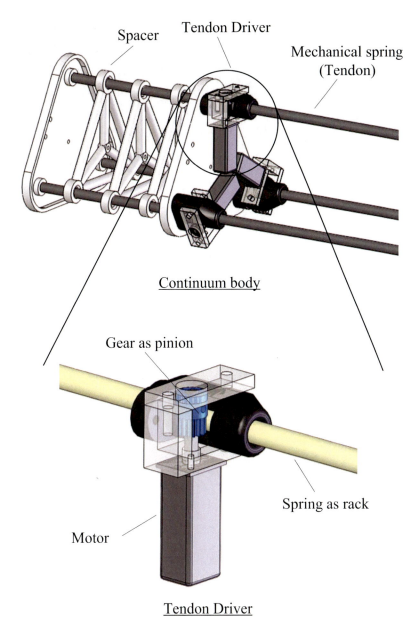

Fig. 4.6 Mechanical design of the proposed continuum body.

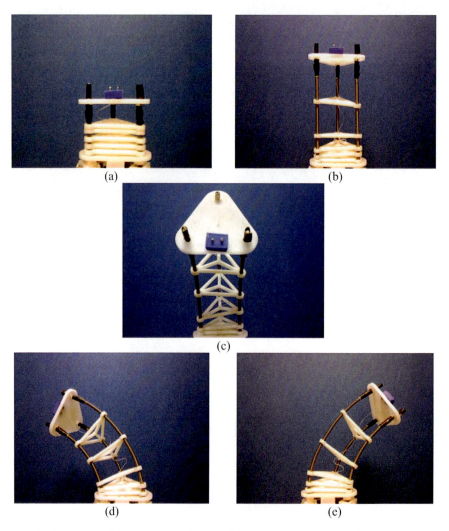

Fig. 4.7 Prototype of the continuum body and illustration of the admissible motions: (a) contraction; (b) extension; (c) forward bending; (d) right bending; (e) left bending.

4.3.3 Semi-passive Joint

To appress the gripper to a gripping surface, the gripper should have a certain turning ability about the y- and z-axes. However, the inherent compliance of the continuum body does not include rotational motion about the y-axis and only affords a limited twisting angle about the z-axis. An additional device is thus needed to provide enough degrees of freedom. To provide the requisite degrees of freedom and maintain the light weight of the robot, a semi-passive joint is developed that comprises a

4.3 Structure

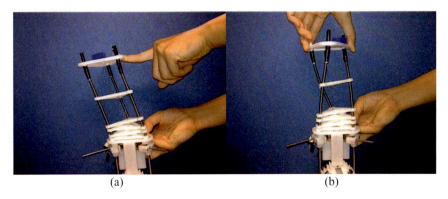

Fig. 4.8 Compliance of the continuum body: (a) shearing and (b) twisting.

passive revolute joint with two DOFs. A passive joint is used rather than an active joint to eliminate the need for complex controls to orient the joints and to reduce the number of actuators, thereby keeping Treebot lightweight. The joint is installed between the front gripper and the continuum body, and can lock and unlock actively. To reduce the number of actuators required, the lock/unlock action is controlled by a linear motor that controls the gripping motion of the front gripper. When the joint is unlocked, the joint can be rotated about the y- and z-axes. When the joint is locked, it actively returns to the initial orientation in which rotation about the y- and z-axes is zero. The locking mechanism is needed to fix the front gripper for exploring purpose.

The semi-passive joint is composed of three parts, Part A, B, and C, as illustrated in Fig. 4.9. Part A is connected to the gripper, Part B is connected to the continuum body, and Part C keeps Part A inside Part B. It can be observed that the joint can freely be rotated in the y- and z-axes only. The range of the twisting angle on the y-axis is ± 45 degrees. There is no mechanical constraint on the twisting angle for the z-axis. However, if the joint turns over ± 60 degrees, the locking system is unable to force the joint to return to the initial orientation. As a result, the angle of twist about the z-axis must be constrained to ± 60 degrees.

At the contact surfaces of Part A and B, there is a convex and a concave triangular cone respectively. The convex triangular cone fits inside the concave triangular cone, and together the cones work as a locking system to return the joint to the initial orientation. The locking mechanism is illustrated in Fig. 4.10. A wire passes through the center of Parts A and B along the z-axis. One end of the wire is fixed on the continuum body, and the other end is connected to the linear motor at the front gripper. The semi-passive joint shares an actuator with the front gripper. When the linear motor extends, the wire pulls Part B close to Part A. The convex and concave triangular cones between the two parts then push together and force the joint to return to the initial orientation.

Fig. 4.11 shows the hardware prototype of the semi-passive joint in an unlocked state. It can be observed that the joint can be twisted about the z-axis. Fig. 4.12 also

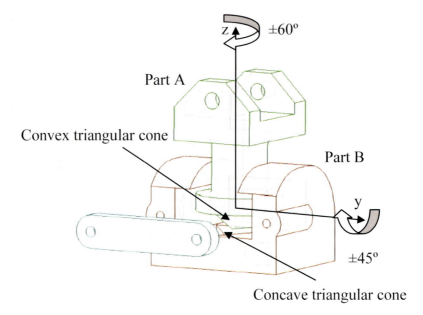

Fig. 4.9 Design of the semi-passive joint.

illustrates the gripper in unlocked and locked state. In Fig. 4.12(a), the front gripper is closed. The semi-passive joint is unlocked, and the front gripper is not in the initial orientation. When the front gripper is opened, the semi-passive joint is locked (Fig. 4.12(b)) and the front gripper returned to the initial orientation.

4.3.4 Sensors

To realize the motions of Treebot and explore the environment, three types of sensors are used, i.e., encoders, tactile sensors, and tilting sensors. Encoders are installed on each tendon of the driving motor to measure the posture of the continuum body. Four tactile sensors which act as tentacles are installed on each gripper to detect the interaction between the gripper and the climbing surface. A triple-axis tilting sensor is also attached to the front gripper to measure the direction of gravity.

4.4 Locomotion

The locomotion of Treebot is similar to that of inchworms, which is a kind of bipedal locomotion. Fig. 4.13 shows a complete climbing gait of the locomotion. It is composed of six climbing procedures. The square colored in grey represents the closed

4.4 Locomotion

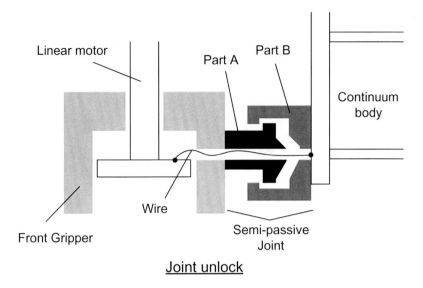

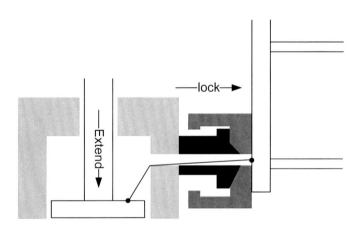

Fig. 4.10 Locking mechanism of the semi-passive joint.

gripper that attaches to the substrate, while the square colored in white represents the opened gripper that detaches from the substrate. The order of the motions illustrated in the figure represents the locomotion of moving forward. The locomotion of moving backward is the same but in reverse order.

Treebot is capable of changing its direction of movement in three-dimensional space by bending the continuum body. This allows Treebot to climb along a curved-shape tree or avoid obstacles, giving Treebot high maneuverability surpassing that

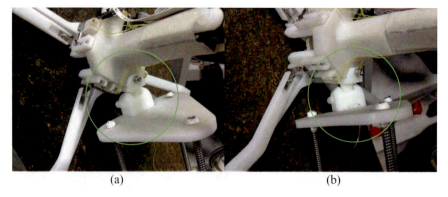

Fig. 4.11 Hardware prototype and the installation position of the semi-passive joint. (a) Initial orientation; (b) Twisted about the z-axis.

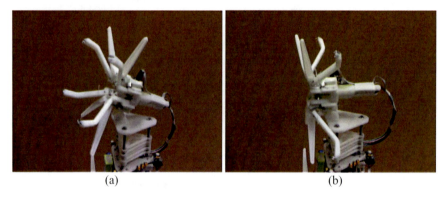

Fig. 4.12 Semi-passive joint: (a) unlocked; (b) locked.

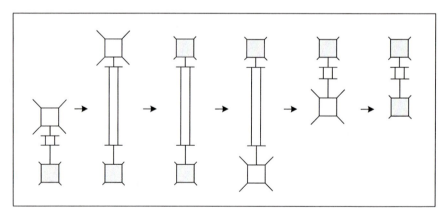

Fig. 4.13 A complete climbing gait of Treebot (moving forward).

of the existing tree-climbing robots. Fig. 4.14 shows a sequence of climbing motions that allow it to avoid an obstacle on a tree. In the motion, Treebot first adjusts the direction of the rear gripper and then climb along that direction to avoid the obstacle. This method is also applicable for turning to another side of a branch or selecting a branch to climb.

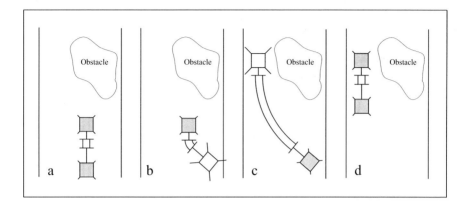

Fig. 4.14 Motions to avoid an obstacle.

4.5 Hardware Prototype

Fig. 4.15 and Fig. 4.16 show the prototypes of Treebot. The first prototype named Treebot is a remote-control robot with no sensor installed. It is used to verify the mechanical design by manual control. The components are mainly made of polyoxymethylene plastic and aluminum alloy to keep the robot in lightweight, and the springs on the continuum body are made of steel. The prototype weighs only 600 grams, which is very light when compared with the existing tree-climbing robots. It is capable of extending by a maximum of 340mm, and has a climbing speed of 22.4cm/min.

The second prototype, Treebot-Auto, is designed to implement the autonomous climbing strategy, and is equipped with several sensors. Several performances such as climbing speed have been improved when compared with the first prototype. The weight of Treebot-Auto, including the battery, is 650 grams, which is slightly heavier than the previous version. The detailed specifications of the prototypes are summarized in Table 4.1.

Fig. 4.15 The first prototype: Treebot.

4.5 Hardware Prototype

Fig. 4.16 The second prototype: Treebot-Auto.

Table 4.1 Specifications of Treebot

Parameters	Treebot	Treebot-Auto
Weight	600grams	650 grams
Height	135mm	135mm
Width	175mm	175mm
Length (minimum)	325mm	370mm
Length of extension	340mm	220mm
Maximal bending curvature	$1/30mm^{-1}$	$1/30mm^{-1}$
Power source	NiMh 4.8V 1000mAh	LiPo 7.4V 800mAh
Runtime	2.5 hours	3 hours
Maximal climb-up speed	26.4cm/min	73.3cm/min
Maximal inclined angle	105°	105°
Adaptable tree diameter	64-452mm	64-452mm

4.6 Energy Consumption

One of the excellent features of Treebot is that it consumes little energy. It can even consume zero energy when it holds on a tree. This is accomplished by the special design of the gripper and the self-locking characteristic of the actuators. The actuators can attach on a position without consuming any energy as the gripping force is generated by the preload force of the springs in the gripper.

This feature make Treebot suitable for lengthy work on trees, such as surveillance. Treebot also consumes little energy when climbing, and can climb continuously for almost 3 hours on a LiPo two-cell 7.4V/800mAh battery, which weighs about 45 grams.

4.7 Accessories

Treebot can be equipped with several accessories, such as a camera and a photovoltaic module as shown in Fig. 4.17 and Fig. 4.18 to enhance its functionality.

The camera can be used to inspect the tree surface or for surveillance. It is fixed on the front gripper. The direction of the camera can be controlled by the continuum body, and hence additional actuators are not required.

A photovoltaic module can be equipped if Treebot is needed to work on a tree for long time such as several days or weeks. As Treebot is designed to work outdoors, renewable solar energy is the best option for providing unlimited energy. This will eliminate the need to replace the battery and enable Treebot to remain and work on a tree independently.

4.8 Control

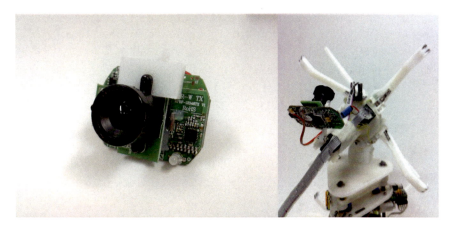

Fig. 4.17 A camera module.

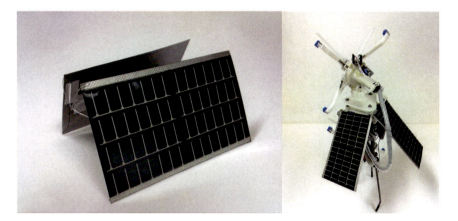

Fig. 4.18 A photovoltaic module.

The wireless camera weighs only 12 grams, and the photovoltaic module just 10.9 grams. As both accessories are lightweight, they will not affect much of the climbing performance.

4.8 Control

4.8.1 Control Architecture

The control architecture is divided into two parts: the ground station and the embedded microcontroller in the robot itself. Fig. 4.19 illustrates the control architecture and the functions of each part, and Fig. 4.20 shows the interface of the ground station. The control mechanism has a master-slave architecture. The ground station works as a master, making decisions and monitoring the state of Treebot. A video signal from Treebot can be displayed on the ground station in real-time. The user can also control Treebot manually through the ground station. The embedded microcontroller works as a slave to implement the motions commanded by the ground station.

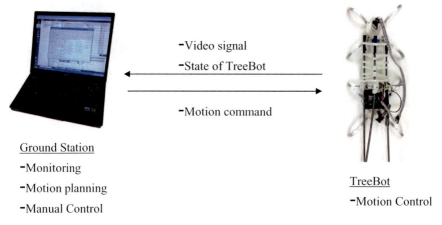

Fig. 4.19 Control architecture of Treebot.

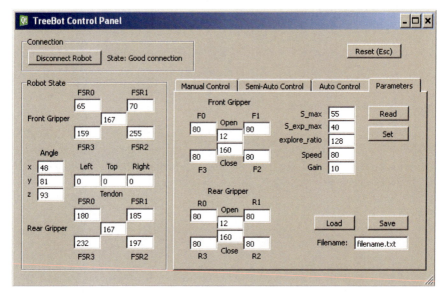

Fig. 4.20 Interface of the ground station.

4.9 Experiments

Decision-making involves complex algorithms that require a high computational power. If the computational work was performed by Treebot itself, then the robot would need a large processor and would consume much more energy, which would make it heavier. Decision-making is thus carried out on the ground station to allow the robot to be lighter.

4.8.2 Manual Control

Treebot can be a remote control robot. The control input of the gripper is simply an on/off command, and it makes the grippers fully open or close. As for the control of the continuum body, since it has three DOFs, three channels of input are needed. One way to control Treebot is to input the length of each spring directly. However, it is not an intuitive method for human manipulation. Humans always have a perspective of the direction of motion when controlling something, i.e., the concept of left, right, front, and back. As a result, to make an intuitive controller, we define three control inputs, i.e., S_{input}, κ^{FB}_{input} and κ^{LR}_{input}. Fig. 4.21 shows the correspondent configuration of the control panel. S_{input} controls the length of the continuum body, κ^{FB}_{input} controls the magnitude of front and back bending while κ^{LR}_{input} controls the magnitude of left and right bending. The concept of front is defined as the direction of positive x-axis, and the concept of left is defined as the direction of positive y-axis. The mapping from the control inputs to the posture of the continuum body are as follows:

$$\begin{bmatrix} S \\ \kappa \\ \phi \end{bmatrix} = \begin{bmatrix} \min\left(\sqrt{{\kappa^{FB}_{input}}^2 + {\kappa^{LR}_{input}}^2}, \kappa_{\max}\right) \\ \mathrm{atan2}\left(\kappa^{LR}_{input}, \kappa^{FB}_{input}\right) \end{bmatrix} \tag{4.1}$$

where $S_{input} \in [0, S_{\max}]$ and $\kappa^{FB}_{input}, \kappa^{LR}_{input} \in [-\kappa_{\max}, \kappa_{\max}]$. S, κ and ϕ are the parameters to describe the posture of the continuum body that is defined in Chapter 6. Once the posture of the continuum manipulator is determined, the length of each spring can be found by (6.1).

4.9 Experiments

Numerous experiments have been conducted to evaluate the performance of Treebot in various aspects, including *generality*, *transition motion*, *turning motion*, *slope climbing* and *payload*.

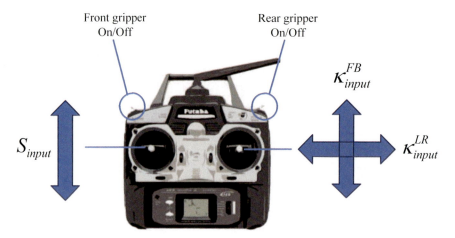

Fig. 4.21 Configuration of the control panel to control Treebot.

4.9.1 Generality

Tree-climbing tests were carried out on thirteen species of trees. The tree species, bark texture, diameter, and the number of trials, and number of successful climbing gaits (as illustrated in Fig. 4.13) are summarized in Table 4.2. Some of the testing environments are also shown in Fig. 4.22. The results show that Treebot performs well on a wide variety of trees. It can be realized that the range of successful climbing diameter of tree is wide, from 64mm to 452mm. However, Treebot did fail to climb several species of trees, including Melaleuca quinquenervia, Cinnamomum camphora, and Bambusa vulgaris var. Striata. Treebot failed to climb on Bambusa vulgaris var. Striata because the surface of this tree is very hard and difficult for the spine on the gripper to penetrate. Melaleuca quinquenervia and Cinnamomum camphora both have peeling bark, and although the gripper could grip them, Treebot fell off as the bark peeled away. These experimental results indicate that Treebot performs well on trees with surfaces that are not particularly hard and on non-peeling bark.

4.9.2 Transition Motion

To verify the maneuverability of Treebot, its transition motion from a trunk to a branch was tested on a Bauhinia blakeana. The diameter of the trunk was 280mm and the slope was about 45 degrees. The diameter of the target gripping branch was 118mm and the slope was about 90 degrees. Some of the transition motions are depicted in Fig. 4.23. It shows that Treebot successfully left the trunk and completely climbed on the branch by bending the continuum body backward to fit the shape. This transition motion takes three climbing steps in three minutes.

4.9 Experiments

Table 4.2 Climbing performance on different species of trees

Tree	Bark texture	Diameter (mm)	No. of steps (Success/Total)
Bombax malabaricum	Rough	452	20/20
Callistemon viminalis	Ridged and furrowed	315	20/20
Delonix regia	Smooth	309	20/20
Bauhinia blakeana	Smooth	80, 207	20/20
Bauhinia variegate	Smooth	258	20/20
Roystonea regia	Shallowly fissured, smooth	352	20/20
Acacia confuse	Smooth	229	20/20
Grevillea robusta	Scaly	159	20/20
Bambusa ventricosa	Smooth	64, 95	20/20
Araucaria heterophylla	Banded	277	20/20
Cinnamomum camphora	Ridged and furrowed, exfoliating	210, 293	13/20
Bambusa vulgaris var. Striata	Smooth, hard	99	1/5
Melaleuca quinquenervia	Sheeting, soft, exfoliating	446	0/5

Another transition motion was tested on a Delonix regia with diameter 310mm as shown in Fig. 4.24. Treebot succeeded to climb from a trunk to a branch at right hand side. This motion took five climbing steps and around five minutes. In Fig. 4.24(c), it can be observed that the continuum body was twisted about 90 degrees along z-axis to make the front gripper appress to the tree surface. It demonstrated that the passive compliance on the continuum body is useful to help Treebot adapts unstructured environment.

4.9.3 Turning Motion

A turning motion has also been performed to evaluate the maneuverability of Treebot. The experiment was carried out on a Bauhinia blakeana trunk with a diameter of 207mm. Fig. 4.25 depicts part of the turning motions. It can be seen that Treebot successfully moved from the front to the back of the trunk. This motion took five

Fig. 4.22 Climbing test on different types of trees: (a) Callistemon viminalis; (b) Bambusa ventricosa; (c) Araucaria heterophylla; (d) Cinnamomum camphora; (e) Bambusa vulgaris var. Striata; (f) Melaleuca quinquenervia.

climbing steps and around five minutes. The compliance of the gripper resulted in successful appression to the tree surface (Fig. 4.25(b), (c) and (d)), allowing Treebot to perform the turning motion successfully.

4.9.4 Slope Climbing

The limitation of the slope that Treebot is capable of climbing is 105 degrees. The details of this limit can be found in Chapter 6. An experiment was conducted to examine the maximum climbing slope of Treebot. It has been implemented on a Bauhinia blakeana with a diameter of 172mm. The climbing angle was about 103 degrees. Part of the climbing motions is depicted in Fig. 4.26, which shows that Treebot was able to climb the slope successfully. There was no over slope climbing effect appeared in the experiment.

4.9 Experiments

Fig. 4.23 Branch transition on a Bauhinia blakeana.

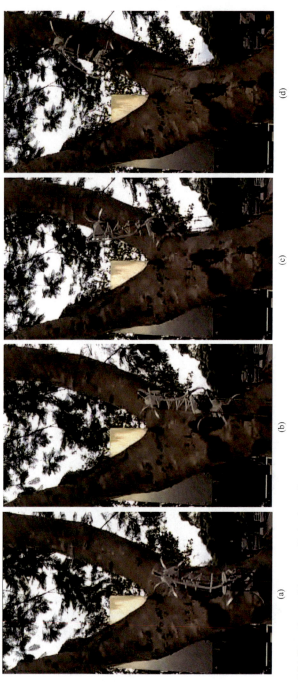

Fig. 4.24 Branch transition on a Delonix regia.

4.9 Experiments

Fig. 4.25 Turning motion on a Bauhinia blakeana.

It was also attempted to make Treebot climb a tree with a slope larger than its climbing limit. As shown in Fig. 4.27, Treebot tried to climb a Bauhinia blakeana with a diameter of 207mm and a climbing angle of about 110 degrees. It can be noticed in Fig. 4.27(b) and (d) that an over-slope climbing effect occurred. This position cannot be adjusted as the compliance of Treebot as insufficiently to compensate for the outward angle. As a result, Treebot cannot climb up further.

4.9.5 Payload

As Treebot is designed to carry equipment up a tree, a payload test was implemented to realize the maximum payload of Treebot. The experiment revealed that Treebot can climb with 1.75kg of extra weight (Fig. 4.28), which is almost three times its own weight.

4.10 Performance Comparison

Table 4.3 makes a clear comparison among Treebot and other tree-climbing robots. The best performance for each item is highlighted. The "/" sign denotes that the data are not available. By neglecting the unknown data, it can be seen that Treebot ranks the best in most aspects except for climbing speed and number of actuators. However, Treebot has only one more actuator than the best performing robot in this respect (Kawasaki's pruning robot), and although it is slower than the fastest climber (RiSE V3), it can still climb faster than the other two robots.

4.11 Summary

In this chapter, the development of a novel tree-climbing robot, Treebot, is presented. Treebot is lightweight and compact. It is composed of a pair of omni-directional tree grippers that allow the robot to hold onto trees and a novel three DOF continuum body for maneuvering. Numerous experiments were conducted to test its performance. The results reveal that the proposed design is capable of climbing a wide variety of trees with a high maneuverability. The range of gripping curvatures is also very wide. It is found that Treebot has excellent climbing performance on the trees with surfaces that are not very hard and do not have easy peeling bark.

One of the original contributions of this work is the application of the extendable continuum mechanism as a maneuvering system for tree climbing. This opens up a new field of applications for the continuum mechanism. Through studies and experiments, it is discovered that the extendable continuum mechanism is highly suitable for tree-climbing application. It gives Treebot has high maneuverability such that the admissible climbing workspace surpasses that of all the state of the art

4.11 Summary

Fig. 4.26 103 degrees slope climbing.

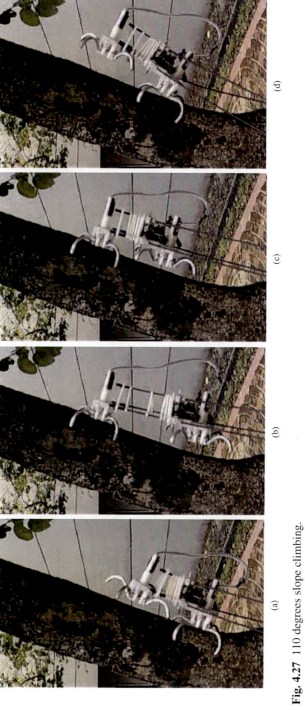

Fig. 4.27 110 degrees slope climbing.

4.11 Summary

Table 4.3 Comparison among the existing tree-climbing robots

	Treebot	Woody [15]	Kawasaki's pruning robot [16]	RiSE V2 [18]	RiSE V3 [19]	TREPA [17]
Weight (kg)	0.65	13.8	15	3.8	5.4	/
Payload capacity (kg)	1.75	/	/	1.5	/	/
Height (mm)	135	310	400	/	/	/
Width (mm)	175	310	400	333	275	/
Length (mm)	370	750	400	583	980	/
Runtime (minutes)	180	/	/	45	30	/
Climbing speed (mm/min)	733	1164	2400	537	12600	333
Adaptable tree diameter (mm)	64-452	100-150	/<120</	/ - ∞	/<250</	/
Number of actuators	5	/	4	12	8	14
Workspace	curved trunk and branches	straight trunk	straight trunk	straight trunk	straight trunk	slightly curved trunk

Fig. 4.28 Climbs with 1.75kg payload.

tree-climbing robots. The inherent compliance of the continuum mechanism also simplifies the control issues and keeps Treebot lightweight.

Another contribution is the development of a miniature omni-directional tree gripper. This unique mechanical design makes the gripper compact and simple to control. It consumes zero energy in static gripping, which enables Treebot to remain on a tree for a long time. The gripper is also able to attach to a wide variety of trees with a wide range of gripping curvatures. This permits Treebot to climb between a large tree trunk and small branches without any change in the gripper settings. On top of that, the gripper settings are optimized analytically to generate the maximum gripping force which detail is discussed in Chapter 5.

Chapter 5
Optimization of the Fastening Force[1]

The proposed tree gripper introduced in Chapter 4 is simple in control. It also implies that it has less controllability. In order to tackle the variant environment in limited controllability, the setting of the gripper should be optimized so as to maximize the fastening force in a wide range of surface curvature. The optimized items include the spine installation angle and the spring force distribution so as to provide optimal spine insert angle and direction of action force over a wide range of surface curvatures. On-tree experiments of the proposed mechanism will be presented at the end of this chapter to evaluate the optimization results.

5.1 Gripping Configuration

The gripper is designed for gripping convex surface. The gripping position can be divided into two parts, that is a closer-half (smaller than 180 degrees gripping angle) and a farer-half (more than 180 degrees gripping angle), as illustrated in Fig. 5.1. The figure also illustrates the simplified gripper model and its parameters. L_c is the length of the claw. w denotes the distance between the revolute joint of the claw and the principal axis of the gripper. h denotes the vertical distance between the revolute joint of the claw and a gripping substrate. r is the radius of the gripping substrate. When the claws are long enough to grip the farer-half of a substrate, a large gripping force can easily be achieved, as just maintaining the claw position is enough to lock the gripper onto the substrate (a force-closure grasp). According to Fig. 5.2(a), to avoid the collision among the claws, the minimal radius of the gripping substrate by interlocking method is,

$$r = \frac{\sqrt{L_c^2 - w^2} - h}{2} \tag{5.1}$$

---

[1] Portions reprinted, with permission from Tin Lun Lam, and Yangsheng Xu, "Mechanical Design of a Tree Gripper for Miniature Tree-Climbing Robots", Proceedings of the IEEE/RSJ International Conference on Intelligent Robots and Systems. ©[2011] IEEE.

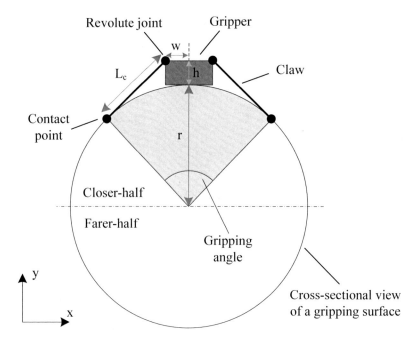

Fig. 5.1 Representation of the gripping position.

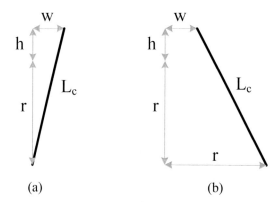

Fig. 5.2 (a) Notations to find out the minimal radius of a substrate by interlocking method. (b) Notations to find out the minimal radius of a substrate that the gripper is capable of gripping the closer-half of the substrate only.

5.2 Generation of the Adhesive Force

If the gripper is to be compact, then the claws cannot be made sufficiently long to always be capable of griping the farer-half of the substrate. In such cases, the gripper must generate sufficient adhesive force by the penetration of spines into a tree surface to avoid being pulled out by the pitch-back moment. According to Fig. 5.2(b), the minimal radius of a gripping substrate that the gripper is only capable of gripping the closer-half is,

$$r = \frac{w - h + \sqrt{L_c^2 - (h+w)^2}}{2} \tag{5.2}$$

As it is aimed to make a miniature robot, the tree gripper is mainly designed and optimized for gripping on the closer-half of a gripping substrate.

5.2 Generation of the Adhesive Force

According to [35], the penetration of a spine can generate both shear force (parallel to the substrate surface) and adhesive force (normal to the substrate surface). The magnitudes of these forces are related to the spine insert angle (θ_i) and the direction of acting force. To determine the force generated by the penetration of the spines, the insert angle must be known. It is assumed that the gripping substrate is in a cylindrical shape with a radius of r, as shown in Fig. 5.3(a). When a claw penetrates the substrate at a certain orientation (σ), the gripping curvature becomes an ellipse. Fig. 5.3(c) shows the notations for the gripping curvature and the gripper parameters. θ_s is the spine installation angle and θ_c is the gripping angle of the claw. To find the spine insert angle, the coordinates of the contact point in a $x' - y'$ frame (x_i', y_i') must first be obtained. This can be achieved by finding the intersection point of the ellipse (E) and the circle (C), which represent the gripping curvature and the motion trajectory of claw respectively:

$$E: \frac{x_i'^2}{a_e^2} + \frac{y_i'^2}{b_e^2} = 1 \tag{5.3}$$

$$C: (x_i' - x_c)^2 + (y_i' - y_c)^2 = r_c^2 \tag{5.4}$$

where $b_e = r$, $a_e = r/\cos\sigma$, $x_c = w$, $y_c = r + h$ and $r_c = L_c$.
Rewrite (5.4):

$$x_i'^2 - 2x_c x_i' + x_c^2 + (y_i' - y_c)^2 - r_c^2 = 0$$

$$\left[x_i'^2 + x_c^2 + (y_i' - y_c)^2 - r_c^2\right]^2 = (2x_c x_i')^2 \tag{5.5}$$

Sub. (5.3) into (5.5):

$$\left[a_e^2\left(1 - \frac{y_i'^2}{b_e^2}\right) + x_c^2 + y_i'^2 - 2y_c y_i' + y_c^2 - r_c^2\right]^2 = 4x_c^2 a_e^2\left(1 - \frac{y_i'^2}{b_e^2}\right) \tag{5.6}$$

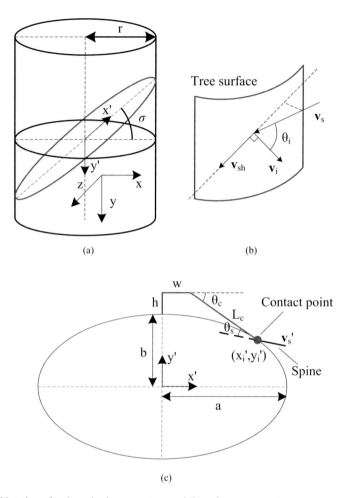

Fig. 5.3 Notations for the gripping curvature and the gripper parameters.

$$\left\{\left[1-\left(\frac{a_e}{b_e}\right)^2\right]y_i'^2 - 2y_c y_i' + \left(x_c^2 + y_c^2 + a_e^2 - r_c^2\right)\right\}^2 = 4x_c^2 a_e^2 \left(1 - \frac{y_i'^2}{b_e^2}\right) \quad (5.7)$$

Let $b_0 = \left[1-\left(\frac{a_e}{b_e}\right)^2\right]$, $b_1 = -2y_c$, and $b_2 = \left(x_c^2 + y_c^2 + a_e^2 - r_c^2\right)$. Eq. (5.7) becomes,

5.2 Generation of the Adhesive Force

$$\left(b_0 y_i'^2 + b_1 y_i' + b_2\right)^2 = 4x_c^2 a_e^2 \left(1 - \frac{y_i'^2}{b_e^2}\right)$$

$$b_0^2 y_i'^4 + 2b_0 b_1 y_i'^3 + \left(2b_0 b_2 + b_1^2\right) y_i'^2 + 2b_1 b_2 y_i' + b_2^2 = 4x_c^2 a_e^2 - 4\left(\frac{x_c a_e}{b_e}\right)^2 y_i'^2$$

$$a_0 y_i'^4 + a_1 y_i'^3 + a_2 y_i'^2 + a_3 y_i' + a_4 = 0 \tag{5.8}$$

where $a_0 = b_0^2$, $a_1 = 2b_0 b_1$, $a_2 = \left[2b_0 b_2 + b_1^2 - 4\left(\frac{x_c a_e}{b_e}\right)^2\right]$, $a_3 = 2b_1 b_2$, and $a_4 = b_2^2 - 4x_c^2 a_e^2$.

The derivation of (5.8) involves finding the roots of a quartic equation which can be solved by Ferrari's method. In that, real roots should be chosen as the solutions. After obtaining y_i', x_i' can be determined by,

$$x_i' = \pm \frac{a_e}{b_e} \sqrt{b_e^2 - y_i'^2} \tag{5.9}$$

Once the intersection point (x_i', y_i') has been found, the spine direction vector $(\mathbf{v}_s')$ can be determined from the following equation according to Fig. 5.4:

$$\mathbf{v}_s' = (-\sin\theta_{sd}, \cos\theta_{sd}) \tag{5.10}$$

where $\theta_{sd} = \theta_c - \theta_s$, $\theta_c = \pi - \theta_a - \theta_b$, $\theta_a = \tan^{-1}\left(\frac{h+r}{w}\right)$, $\cos\theta_b = \frac{L_b^2 + L_c^2 - L_i^2}{2 L_b L_c}$, $L_b^2 = (h+r)^2 + w^2$, and $L_i^2 = x_i'^2 + y_i'^2$.

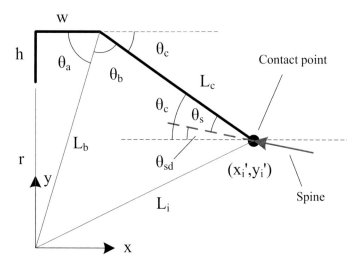

Fig. 5.4 Notations for defining the spine direction vector.

To transform the coordinates from the $x' - y'$ frame to the $x - y - z$ frame, a rotation matrix about the y-axis ($Rot_y(\sigma)$) is used. The spine direction vector $\mathbf{v}_s$ and the coordinates of intersection $\begin{bmatrix} x_i & y_i & z_i \end{bmatrix}$ in the $x - y - z$ frame becomes:

$$\mathbf{v}_s = Rot_y(\sigma)\mathbf{v}_s' \tag{5.11}$$

$$[x_i\ y_i\ z_i]^T = Rot_y(\sigma)[x_i'\ y_i'\ 0]^T \tag{5.12}$$

As the gripping substrate is approximated as a cylinder, the normal vector of the intersection point becomes:

$$\mathbf{v}_i = \left(\frac{x_i}{\sqrt{x_i^2+y_i^2}}, \frac{y_i}{\sqrt{x_i^2+y_i^2}}\right) \tag{5.13}$$

Referring to Fig. 5.3, the spine insert angle θ_i, i.e., the angle between $\mathbf{v}_s$ and $\mathbf{v}_i$ can be obtained from:

$$\theta_i = \cos^{-1}(-\mathbf{v}_i \cdot \mathbf{v}_s) \tag{5.14}$$

As shown in Fig. 5.3(b), the direction of the adhesive force ($\mathbf{v}_{ad}$) is equivalent to $\mathbf{v}_i$, and the direction of the shear force ($\mathbf{v}_{sh}$) can be found by:

$$\mathbf{v}_{sh} = \mathbf{v}_i \times \mathbf{v}_s \times \mathbf{v}_i \tag{5.15}$$

The total force generated by spine penetration is then:

$$\mathbf{F} = f_{ad}\mathbf{v}_{ad} + f_{sh}\mathbf{v}_{sh} \tag{5.16}$$

where f_{ad} and f_{sh} are the magnitude of the adhesive force and shear force respectively.

According to the experimental results of [35], the relationships among the spine insert angle, adhesive force, and shear force are illustrated in Fig. 5.5. The pull-in force (the force that pulls the gripper towards the gripping surface) to compensate for the pitch-back moment is contributed by the adhesive force and shear force of the spine along the z-axis, i.e., F_z where $\mathbf{F} = [F_x\ F_y\ F_z]$.

5.3 Optimization of the Spine Installation Angle

To generate maximal pull-in force in a range of gripping curvatures (K), the spine installation angle (θ_s) must be optimized. As the gripper is designed for omnidirectional gripping about its principal axis as illustrated in Fig. 4.3, the gripping force in different gripping orientations must also be considered. As the claws are symmetrical in both x- and y-axes, the three gripping orientations shown in Fig. 5.6 are equivalent to sixteen gripping orientations that approximately covers most of the possible solution. As a result, the optimization is based on these three gripping orientations and a range of gripping curvatures.

The pull-in force of the gripper is contributed by the force of all claws. The directions of the claws of the three gripping orientations are composed of $\sigma = 0, \frac{\pi}{8}, \frac{\pi}{4}, \frac{3\pi}{8}, \frac{\pi}{2} = \sigma_a, \sigma_b, \sigma_c, \sigma_d, \sigma_e$ where Orientation 1 is composed of $2\sigma_a + 2\sigma_e$;

5.3 Optimization of the Spine Installation Angle

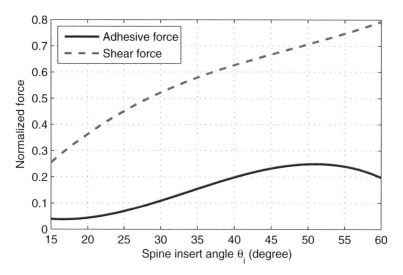

Fig. 5.5 Relationships among the spine insert angle, adhesive force, and shear force [35].

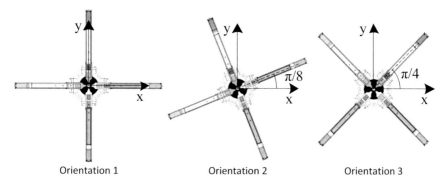

Fig. 5.6 Representative gripping orientations.

Orientation 2 is composed of $4\sigma_c$; and Orientation 3 is composed of $2\sigma_b + 2\sigma_d$. Fig. 5.7 to Fig. 5.11 shows the relationships among the spine installation angle, curvature of gripping surface (K) and pull-in force (F_i, $i = a, b, c, d, e$), at different claw directions with the gripper parameters: $w = 20$mm, $L_c = 100$mm, $h = 25$mm and the range of tree radius: 64.5mm $< r <$ 500mm. The minimum target radius 64.5mm is obtained by (5.2) at which the gripper cannot grip the farer-half of the substrate.

The normalized gripping force generated by different gripping orientations of the gripper can be obtained by summing the results according to the combination of claw directions and then divided by four as illustrated in Fig. 5.12 to Fig. 5.14.

As the gripper is intended to be used omnidirectionally, we must find an optimal installation angle that generate largest force in all gripping orientations. To help determining the optimal installation angle, the average pull-in force generated by three

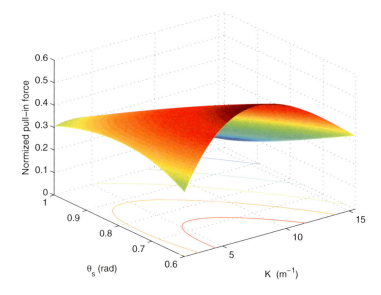

Fig. 5.7 Relationships among the spine installation angle, the curvature of the gripping surface and the normalized pull-in force at $\sigma = 0$.

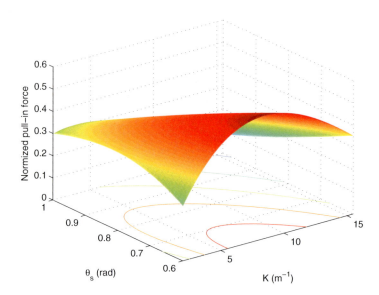

Fig. 5.8 Relationships among the spine installation angle, the curvature of the gripping surface and the normalized pull-in force at $\sigma = \pi/8$.

5.3 Optimization of the Spine Installation Angle

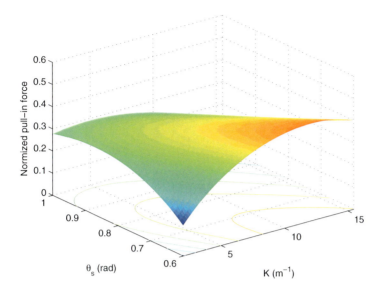

Fig. 5.9 Relationships among the spine installation angle, the curvature of the gripping surface and the normalized pull-in force at $\sigma = \pi/4$.

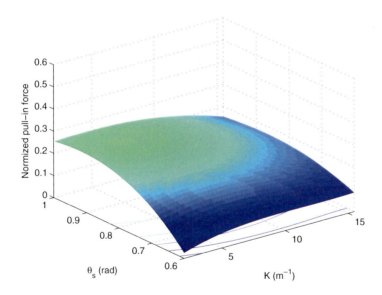

Fig. 5.10 Relationships among the spine installation angle, the curvature of the gripping surface and the normalized pull-in force at $\sigma = 3\pi/8$.

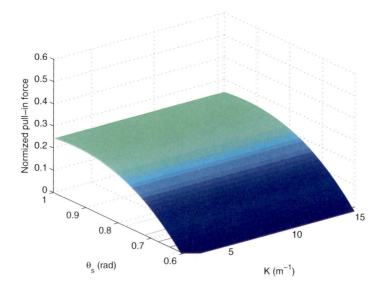

Fig. 5.11 Relationships among the spine installation angle, the curvature of the gripping surface and the normalized pull-in force at $\sigma = \pi/2$.

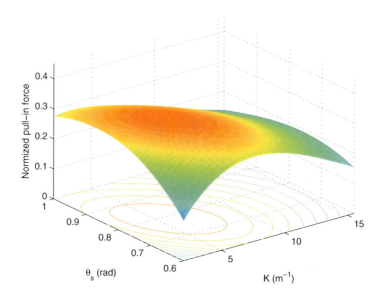

Fig. 5.12 Normalized pull-in force generated by the gripping posture Orientation 1 in different spine installation angles and gripping curvatures.

5.3 Optimization of the Spine Installation Angle   65

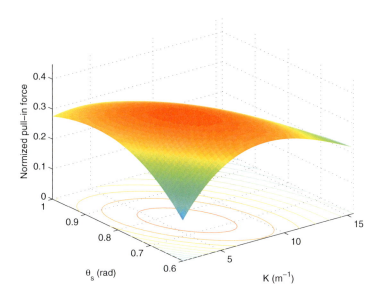

Fig. 5.13 Normalized pull-in force generated by the gripping posture Orientation 2 in different spine installation angles and gripping curvatures.

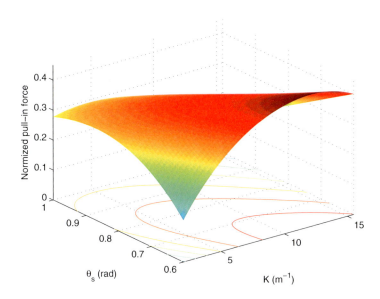

Fig. 5.14 Normalized pull-in force generated by the gripping posture Orientation 3 in different spine installation angles and gripping curvatures.

gripping orientations in different spine installation angles and gripping curvatures are obtained and illustrated in Fig. 5.15. As seen in the figure, the minimal pull-in force is greatest when $\theta_s = 0.76$rad. This is thus defined as the optimal spine installation angle and it is adopted in the gripper design. Fig. 5.16 shows the normalized pull-in force generated by the gripper in different orientations and different gripping curvatures when $\theta_s = 0.76$rad. In the figure, $O1$, $O2$, and $O3$ represent Orientation 1, 2, and 3 respectively. It can be realized from the figure that Orientation 3 generates the largest pull-in force by this optimal setting. Orientation 3 is thus used as the default setting for Treebot.

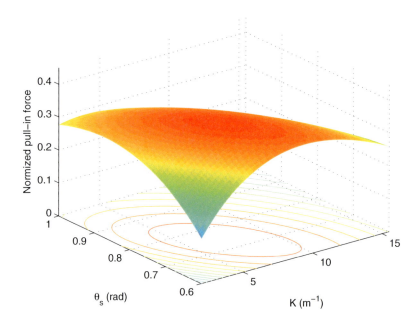

Fig. 5.15 Normalized pull-in force generated by the average of the gripping orientations in different spine installation angles and gripping curvatures.

5.4 Generation of the Directional Penetration Force

In [35], it is mentioned that the optimal direction of the force acting on a spine F_p is equal to the spine insert angle. Pushing a claw into a substrate with a desired directional force usually requires two actuators. RiSE V2 [18], for example, uses two active joints to accomplish this task. One actuator provides a pushing force toward a

5.4 Generation of the Directional Penetration Force

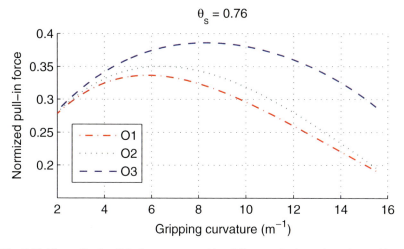

Fig. 5.16 Normalized pull-in force generated by different gripping orientations with optimal spine installation angle $\theta_s = 0.76$ rad.

surface, and the other pulls the spine toward the central axis of the robot body. Wile [54] proposed a mechanism that provides the desired directional force using one actuator to make the gripper lighter. However, this is only applicable on flat surfaces.

In view of this, a preloaded two-bar linkage mechanism is proposed to generate the desired directional force, which requires only one actuator and is able to adapt to irregular surfaces. The notations for the mechanism of the claw are shown in Fig. 5.17. L_1 denotes the length of link AB (Phalanx 1), and L_2 denotes the length of link BC (Phalanx 2). Link AC is the simplified claw introduced in Fig. 5.4. Joints A and B are passive revolute joints installed with a pre-compressed mechanical spring. The torque generated by spring on joints A and B are denoted as τ_1 and τ_2 respectively. The turning angle of Joint B is limited such that the distance between A and C does not exceed L_c.

To analyze the direction of the acting force, links AB and BC are divided as shown in Fig. 5.18. It is assumed that in this position, point C penetrates the substrate. Hence, it can be assumed that points A and C are fixed revolute joints. The

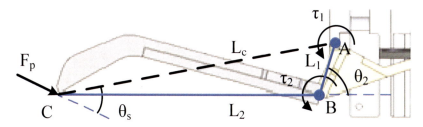

Fig. 5.17 Notations for the mechanism of the claw.

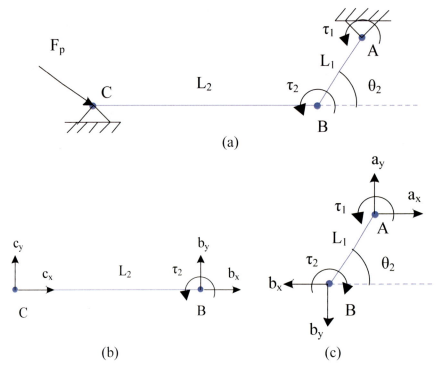

Fig. 5.18 Free body diagrams of links *AB* and *BC*.

free body diagrams of links *AB* and *BC* are constructed as shown in Fig. 5.18(b) and (c).

From Fig. 5.18(b), the equilibrium equations can be divided as:

$$c_x + b_x = 0 \tag{5.17}$$
$$c_y + b_y = 0 \tag{5.18}$$
$$\tau_2 - c_y L_2 = 0 \tag{5.19}$$

From Fig. 5.18(c), the equilibrium equations can be divided as:

$$a_x + b_x = 0 \tag{5.20}$$
$$a_y + b_y = 0 \tag{5.21}$$
$$\tau_1 - \tau_2 + L_1(b_y \cos\theta_2 - b_x \sin\theta_2) = 0 \tag{5.22}$$

Sub. (5.17), (5.18), and (5.19) into (5.22):

$$c_x = \frac{\tau_2 - \tau_1 + \tau_2 \frac{L_1}{L_2} \cos\theta_2}{L_1 \sin\theta_2} \tag{5.23}$$

Divide (5.19) by (5.23):

$$\frac{c_y}{c_x} = \frac{L_1 \sin\theta_2}{L_2\left(1 - \frac{\tau_1}{\tau_2} + \frac{L_1}{L_2}\cos\theta_2\right)} \tag{5.24}$$

The angle of the pushing force of the spine F_p is equal to the spine insert angle, that is,

$$\tan(\angle ACB - \theta_s) = \frac{c_y}{c_x} = \frac{L_1 \sin\theta_2}{L_2\left(1 - \frac{\tau_1}{\tau_2} + \frac{L_1}{L_2}\cos\theta_2\right)}$$

where $\angle ACB = \tan^{-1}\frac{L_1 \sin\theta_2}{L_2 + L_1 \cos\theta_2}$. It implies,

$$\frac{\tau_1}{\tau_2} = 1 + \frac{L_1}{L_2}\left[\cos\theta_2 - \frac{\sin\theta_2}{\tan(\angle ACB - \theta_s)}\right] \tag{5.25}$$

With the gripper parameters: $L_1 = 20$mm, $L_2 = 85$mm, the torque ratio between joints A and B (τ_1/τ_2) should be around 1.4 to generate the appropriate angle of pushing force.

5.5 Experimental Results

Numerous experiments have been implemented on a variety of trees to evaluate the performance of the omni-directional tree gripper. In the experiments, the gripper first gripped a tree without any external force being applied. An external pull-out force was then applied normal to the gripping surface to test how much force was needed to pull the gripper out of the tree. The maximum pull-out force was limited to 40N to avoid breaking the gripper. Eighteen types of trees with different surface curvatures were tested. Fig. 5.19 shows some of the tested trees. The curvature of the trees, bark textures, and the maximum pull-out force with different gripping orientations are summarized in Table 5.1. In the table, $O1$, $O2$, and $O3$ represent Orientations 1, 2, and 3 as introduced in Fig. 5.6 respectively. The curvature of a tree (K) is obtained by:

$$K = \frac{1}{D/2} \tag{5.26}$$

where D is the diameter of a tree.

Table 5.1 shows that on the first ten types of trees (No. 1-10), the performance was excellent. The gripper was capable of generating over 40N of pull-in force in any gripping orientation. However, the results also reveal that the gripper does not work well on some types of trees, and particularly those with bark that peels off easily. In such cases, when a large pull-out force was applied, the gripper was pulled out as the bark peeled off (No. 11-15). Furthermore, for soft trees the pull-out force broke the bark (No. 16-18).

Fig. 5.19 Experiments on different types of trees: (a) Bauhinia variegata var. candida; (b) Roystonea regia; (c) Taxodium distichum; (d) Cinnamomum camphora; (e) Khaya senegalensis; (f) Eucalyptus citriodora.

The experimental results indicate that on most of the trees, the maximum pull-in force of the gripper in all gripping orientations is similar that matches the analytical results reported in the previous section. The only exception is tree No. 13. This is because the bark of this trees peels off easily and its surface is not smooth, but rather has many vertical grooves (see Fig. 5.19(b)). Orientation 1 is better in this scenario, as it creates a pair of claws oriented perpendicular to the vertical groove, which allows the claws to penetrate deeper into the tree to generate a larger force.

As mentioned in the previous section, the gripping curvature affects the pull-in force of the gripper. This phenomenon is clearly demonstrated by the experimental result for tree No. 10, where the generated pull-in force with a $6.2 m^{-1}$ surface curvature is larger than that with a $4.8 m^{-1}$ surface curvature. However, the result for tree No. 11 does not match the analytical result (Fig. 5.16). This is because the tree with a $6.0 m^{-1}$ surface curvature was older, and its bark will be peeled off easily, whereas the tree with a $11.2 m^{-1}$ surface curvature was younger and its bark will not be peeled off easily.

In the experimental results, and especially those for trees No. 16-18, it is clear that using Orientation 3 generates the largest pull-in force, which matches the analytical results.

5.5 Experimental Results

Table 5.1 Maximum pull-in force on different species of trees

No.	Trees	Bark texture	K (m^{-1})	Pull-in force (N) O1	O2	O3
1	Bombax malabaricum	Rough	4.4	>40	>40	>40
2	Acacia confuse	Smooth	5.6	>40	>40	>40
3	Ficus microcarpa	Smooth	4.6	>40	>40	>40
4	Livistona chinensis	Fissured	8.4	>40	>40	>40
5	Callistemon viminalis	Ridged and Furrowed	6.3	>40	>40	>40
6	Bauhinia variegata var. candida	Smooth	8.1	>40	>40	>40
7	Bauhinia variegate	Smooth	8.8	>40	>40	>40
8	Araucaria heterophylla	Banded	7.2	>40	>40	>40
9	Bauhinia blakeana	Smooth	6.7	>40	>40	>40
10	Roystonea regia	Smooth, shallowly fissured	6.2	>40	>40	>40
			4.8	15	15	20
11	Taxodium distichum	Fibrous, exfoliating	11.2	29	30	30
			6.0	12	10	10
12	Casuarina equisetifolia	Slightly exfoliating	6.9	11	13	12
13	Cinnamomum camphora	Ridged and furrowed, exfoliating	5.2	20	12	5
14	Khaya senegalensis	Blocky, exfoliating	4.0	10	10	10
15	Melaleuca quinquenervia	Sheeting, soft, exfoliating	4.5	5	5	5
16	Delonix regia	Smooth	6.7	24	24	25
17	Mangifera indica	Shallowly fissured	4.1	20	22	25
18	Eucalyptus citriodora	Smooth, soft	4.1	18	16	20

5.6 Summary

In summary, the mechanism of the fastening force generation of the tree gripper and the relationship between the gripper and the gripping substrate have been discussed in this chapter. The optimization of the gripper in terms of torque distribution and spine installation angle has also been conducted and presented to provide a large gripping force over a wide range of curvatures. In addition, numerous experiments have been carried out to evaluate the performance of the tree gripper. The experimental results reveal that the proposed gripper performs well on a wide range of trees. However, the gripping performance strongly depends on the properties of the tree surface. The gripper works well only on trees with bark that does not peel off or break easily.

Chapter 6
Kinematics and Workspace Analysis

6.1 Kinematic Analysis

6.1.1 Configuration of Treebot

Fig. 6.1 shows the configuration of Treebot. In the notations, the superscripts r and f denote the front and rear gripper frame respectively. l_f and l_r represent the distance from the end of the continuum body to the center of the front and rear gripper respectively. h_g denotes the distance between the base of the gripper and the continuum body. The reference frames for the front and rear grippers are also illustrated in the figure. The direction of a gripper refers to the direction along the positive z-axis, where a normal direction refers to the direction toward the positive x-axis.

6.1.2 Kinematics of the Continuum Body

Jones [30] developed a kinematic model of a continuum type manipulator. It formulates the mapping between the posture (S, κ, ϕ) and the input coordinates (l_1, l_2, l_3). S, κ, and ϕ denote the length, curvature, and bending direction of the virtual tendon, respectively. The virtual tendon represents the centerline of the continuum manipulator. l_i denotes the length of each tendon, and d is the distance between the tendons and the virtual tendon. Fig. 6.2 shows the notation used to represent the parameters of each tendon and the position of the virtual tendon.

The kinematic model is also applicable to the proposed continuum body. According to Jones [30], the forward and inverse kinematics in our convention can be defined as:

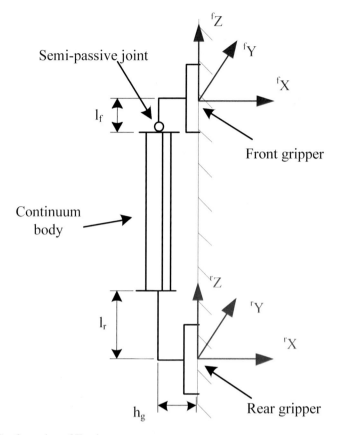

Fig. 6.1 Configuration of Treebot.

Inverse Kinematics $(l_1, l_2, l_3) \leftarrow f(S, \kappa, \phi)$:

$$\begin{bmatrix} l_1 \\ l_2 \\ l_3 \end{bmatrix} = S \begin{bmatrix} 1 + d\kappa \cos \phi \\ 1 - \kappa d \sin\left(\frac{\pi}{6} - \phi\right) \\ 1 - \kappa d \sin\left(\frac{\pi}{6} + \phi\right) \end{bmatrix} \qquad (6.1)$$

Forward Kinematics $(S, \kappa, \phi) \leftarrow f(l_1, l_2, l_3)$:

$$\begin{bmatrix} S \\ \kappa \\ \phi \end{bmatrix} = \begin{bmatrix} \frac{l_1 + l_2 + l_3}{3} \\ 2\frac{\sqrt{l_1^2 + l_2^2 + l_3^2 - l_1 l_2 - l_2 l_3 - l_1 l_3}}{d(l_1 + l_2 + l_3)} \\ \cot^{-1}\left(-\frac{\sqrt{3}}{3} \frac{l_3 + l_2 - 2l_1}{l_2 - l_3}\right) \end{bmatrix} \qquad (6.2)$$

where $\kappa = 1/r$ and $S = r\theta$.

In addition, the mapping between the posture (S, κ, ϕ) and the Cartesian coordinates at the end point (x_t, y_t, z_t) are defined as:

6.1 Kinematic Analysis

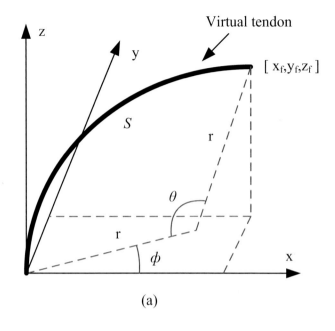

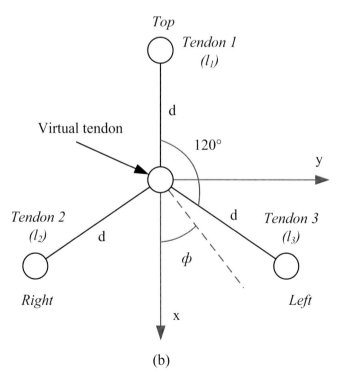

Fig. 6.2 Notations for defining the position and parameters of the continuum manipulator.

$(S, \kappa, \phi) \leftarrow f(x_t, y_t, z_t)$:

$$\begin{bmatrix} S \\ \kappa \\ \phi \end{bmatrix} = \begin{bmatrix} \frac{x_t'^2 + z_t^2}{x_t'} \tan^{-1} \frac{x_t'}{z_t} \\ \frac{2x_t'}{x_t'^2 + z_t^2} \\ \tan^{-1} \frac{y_t}{x_t} \end{bmatrix} \quad (6.3)$$

where $x_t' = x_t \cos\phi + y_t \sin\phi$.

$(x_t, y_t, z_t) \leftarrow f(S, \kappa, \phi)$:

$$\begin{bmatrix} x_t \\ y_t \\ z_t \end{bmatrix} = \frac{1}{\kappa} \begin{bmatrix} [1 - \cos(\kappa S)] \cos\phi \\ [1 - \cos(\kappa S)] \sin\phi \\ \sin(\kappa S) \end{bmatrix} \quad (6.4)$$

The detailed derivation of the equations can be found in Appendix A.1.

6.1.3 Kinematics of Treebot

To formulate the kinematics of Treebot, l_f and l_r must be considered. As a result, the kinematics of Treebot is developed by extending (6.1) and (6.2).

In view of the rear gripper frame as shown in Fig. 6.3(a), the mapping between the end point (front gripper) and the posture of Treebot are formulated as, $({}^r x_f, {}^r y_f, {}^r z_f) \leftarrow f(S, \kappa, \phi)$:

$$\begin{bmatrix} {}^r x_f \\ {}^r y_f \\ {}^r z_f \end{bmatrix} = \begin{bmatrix} \left(\frac{1}{\kappa}[1 - \cos(\kappa S)] + l_f \sin(\kappa S)\right) \cos\phi \\ \left(\frac{1}{\kappa}[1 - \cos(\kappa S)] + l_f \sin(\kappa S)\right) \sin\phi \\ \frac{1}{\kappa} \sin(\kappa S) + l_f \cos(\kappa S) + l_r \end{bmatrix} \quad (6.5)$$

$(S, \kappa, \phi) \leftarrow f({}^r x_f, {}^r y_f, {}^r z_f)$:

$$\begin{bmatrix} S \\ \kappa \\ \phi \end{bmatrix} = \begin{bmatrix} \frac{1}{\kappa} \tan^{-1}\left(\frac{2^r \hat{x}_f ({}^r \hat{z}_f + l_f)}{({}^r \hat{z}_f + l_f)^2 - {}^r \hat{x}_f^2}\right) \\ \frac{2^r \hat{x}_f}{{}^r \hat{x}_f^2 + {}^r \hat{z}_f^2 - l_f^2} \\ \tan^{-1} \frac{{}^r y_f}{{}^r x_f} \end{bmatrix} \quad (6.6)$$

where ${}^r \hat{x}_f = {}^r x_f \cos\phi + {}^r y_f \sin\phi$ and ${}^r \hat{z}_f = {}^r z_f - l_r$.

In view of the front gripper frame as shown in Fig. 6.3(b), the mapping between the end point (rear gripper) and the posture of Treebot are formulated as, $({}^f x_r, {}^f y_r, {}^f z_r) \leftarrow f(S, \kappa, \phi)$:

$$\begin{bmatrix} {}^f x_r \\ {}^f y_r \\ {}^f z_r \end{bmatrix} = \begin{bmatrix} \left(\frac{1}{\kappa}[1 - \cos(\kappa S)] + l_r \sin(\kappa S)\right) \cos\phi \\ \left(\frac{1}{\kappa}[1 - \cos(\kappa S)] + l_r \sin(\kappa S)\right) \sin\phi \\ -\left(\frac{1}{\kappa} \sin(\kappa S) + l_r \cos(\kappa S) + l_f\right) \end{bmatrix} \quad (6.7)$$

6.1 Kinematic Analysis

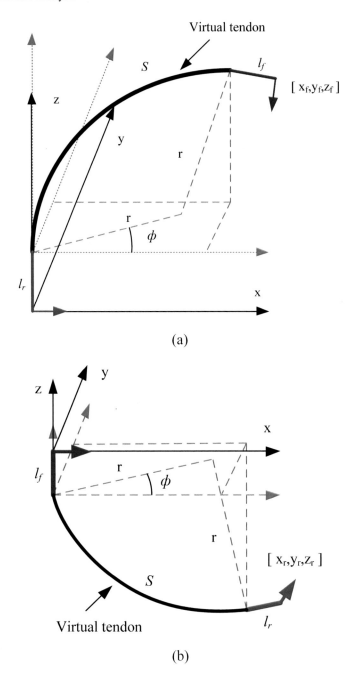

Fig. 6.3 Notations for defining the kinematics of Treebot.

$(S, \kappa, \phi) \leftarrow f\left({}^f x_r, {}^f y_r, {}^f z_r\right)$:

$$\begin{bmatrix} S \\ \kappa \\ \phi \end{bmatrix} = \begin{bmatrix} \frac{1}{\kappa} \tan^{-1}\left(\frac{2{}^f \hat{x}_r \left({}^f \hat{z}_r + l_r\right)}{\left({}^f \hat{z}_r + l_r\right)^2 - {}^f \hat{x}_r^2} \right) \\ \frac{2{}^f \hat{x}_r}{{}^f \hat{x}_r^2 + {}^f \hat{z}_r^2 - l_r^2} \\ \tan^{-1} \frac{{}^f y_r}{{}^f x_r} \end{bmatrix} \quad (6.8)$$

where ${}^f \hat{x}_r = {}^f x_r \cos\phi + {}^f y_r \sin\phi$ and ${}^f \hat{z}_r = -{}^f z_r - l_f$.

On top of that, the transformation of the direction vector between the front and rear gripper frame can be achieved by,

$${}^f \mathbf{v} = Rot_z(\phi) Rot_y(-\theta) Rot_z(-\phi) {}^r \mathbf{v} \quad (6.9)$$

$${}^r \mathbf{v} = Rot_z(\phi) Rot_y(\theta) Rot_z(-\phi) {}^f \mathbf{v} \quad (6.10)$$

The detailed derivation of the equations can be found in Appendix A.2.

6.1.4 Tree Model

Fig. 6.4 illustrates the parameters and coordinate relationship between Treebot and a tree model. The geometry of a segment of tree is modeled as a straight or curved cylinder. The radius of the tree R_{tree}, length of the tree segment S_{tree}, bending direction ϕ_{tree}, and bending curvature κ_{tree} are used to represent the shape of tree. In that, S_{tree}, ϕ_{tree}, and κ_{tree} represent the shape of the centerline of the tree model (similar to the concept of virtual tendon). The target position of the front gripper is defined by the angle of change θ_t and the length of centerline S_t as shown in Fig. 6.4(a). The distance between the continuum body and the tree surface is defined as h_g. The center of the rear gripper is located at the origin of the tree frame (${}^T x - {}^T y - {}^T z$). According to (6.5) and (6.10), the direction of growth of the tree ${}^T \mathbf{v}_t$, the normal vector ${}^T \mathbf{n}_t$, and the coordinates ${}^T \mathbf{P}_t$ at the target position in tree frame can be obtained by:

$${}^T \mathbf{v}_t = Rot_z(\phi_{tree}) Rot_y(\theta_{tree}) Rot_z(-\phi_{tree}) \begin{bmatrix} 0 \\ 0 \\ 1 \end{bmatrix} \quad (6.11)$$

$${}^T \mathbf{n}_t = Rot_z(\phi_{tree}) Rot_y(\theta_{tree}) Rot_z(-\phi_{tree}) Rot_z(\theta_t) \begin{bmatrix} -1 \\ 0 \\ 0 \end{bmatrix} \quad (6.12)$$

$${}^T \mathbf{P}_t = \frac{1}{\kappa_{tree}} \begin{bmatrix} [1 - \cos\theta_{tree}] \cos\phi_{tree} \\ [1 - \cos\theta_{tree}] \sin\phi_{tree} \\ \sin\theta_{tree} \end{bmatrix} + (h_g + R_{tree}) \left(\begin{bmatrix} 1 \\ 0 \\ 0 \end{bmatrix} - {}^T \mathbf{n}_t \right) \quad (6.13)$$

where $\theta_{tree} = \kappa_{tree} S_t$.

6.1 Kinematic Analysis

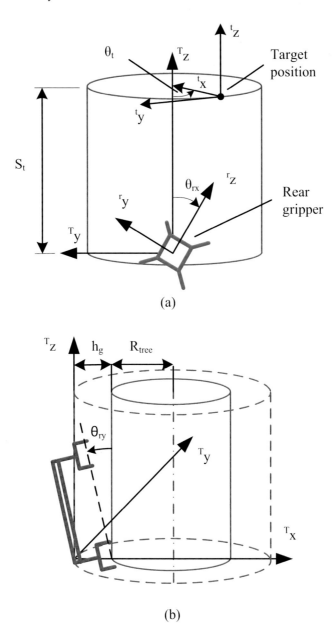

Fig. 6.4 Relationship between the rear gripper and the tree model.

Similar to that of (6.9) and (6.10), the transformation of the direction vector between the tree frame ($^T x - {}^T y - {}^T z$) and the target position frame (${}^t x - {}^t y - {}^t z$) as illustrated in Fig. 6.4(a) can be achieved by,

$$^T \mathbf{v} = Rot_z(\phi_{tree}) Rot_y(\theta_{tree}) Rot_z(-\phi_{tree}) Rot_z(\theta_t)\, {}^t \mathbf{v} \qquad (6.14)$$

$$^t \mathbf{v} = Rot_z(-\theta_t) Rot_z(\phi_{tree}) Rot_y(-\theta_{tree}) Rot_z(-\phi_{tree})\, {}^T \mathbf{v} \qquad (6.15)$$

The direction of growth, normal vector, and the coordinate of the target position can be transformed to the rear gripper frame (${}^r x - {}^r y - {}^r z$) by,

$$^r \mathbf{v}_t = Rot_y(-\theta_{ry}) Rot_x(-\theta_{rx})\, {}^T \mathbf{v}_t \qquad (6.16)$$

$$^r \mathbf{n}_t = Rot_y(-\theta_{ry}) Rot_x(-\theta_{rx})\, {}^T \mathbf{n}_t \qquad (6.17)$$

$$^r \mathbf{P}_t = Rot_y(-\theta_{ry}) Rot_x(-\theta_{rx})\, {}^T \mathbf{P}_t \qquad (6.18)$$

where θ_{rx} and θ_{ry} denote the angles between the tree and the rear gripper frame, as illustrated in Fig. 6.4.

According to (6.9), the direction of growth and the normal vector at the front gripper frame can be determined by,

$$^f \mathbf{v}_t = \begin{bmatrix} {}^f v_{t\_x} \\ {}^f v_{t\_y} \\ {}^f v_{t\_z} \end{bmatrix} = Rot_z(\phi_B) Rot_y(-\kappa_B S_B) Rot_z(-\phi_B)\, {}^r \mathbf{v}_t \qquad (6.19)$$

$$^f \mathbf{n}_t = \begin{bmatrix} {}^f n_{t\_x} \\ {}^f n_{t\_y} \\ {}^f n_{t\_z} \end{bmatrix} = Rot_z(\phi_B) Rot_y(-\kappa_B S_B) Rot_z(-\phi_B)\, {}^r \mathbf{n}_t \qquad (6.20)$$

where S_B, κ_B, and ϕ_B are the length, bending curvature and bending direction of the continuum body.

6.2 Workspace Analysis

By using the proposed continuum mechanism, Treebot is capable of reaching any position in 3D space theoretically. Fig. 6.5 illustrates some of the reachable positions of the continuum body. The arrows in the figure represent the initial direction of the front gripper. However, the reachable workspace of the continuum body is not equivalent to the climbable workspace of Treebot as the gripper works directionally. In addition, the physical constraints of the continuum body must be considered.

6.2 Workspace Analysis

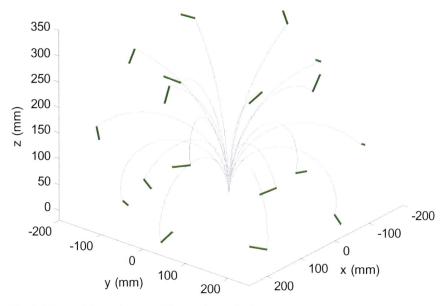

Fig. 6.5 Reachable workspace of the continuum body.

6.2.1 Physical Constraints

6.2.1.1 Maximum Length of Extension

In theory, the tendon-driving mechanism can extend infinitely with infinite length of tendons. In practice, it is impossible to have a tendon of an infinite length. The length of tendons is thus restricted in certain length. Let the maximum length of all tendons be l_{max}. According to (6.1), the length of virtual tendon S is restricted as,

$$S \begin{bmatrix} 1 + d\kappa \cos \phi \\ 1 - \kappa d \sin\left(\frac{\pi}{6} - \phi\right) \\ 1 - \kappa d \sin\left(\frac{\pi}{6} + \phi\right) \end{bmatrix} \leq \begin{bmatrix} l_{max} \\ l_{max} \\ l_{max} \end{bmatrix} \qquad (6.21)$$

It implies,

$$S \leq \begin{bmatrix} \frac{l_{max}}{1+d\kappa\cos\phi} \\ \frac{l_{max}}{1-\kappa d \sin\left(\frac{\pi}{6}-\phi\right)} \\ \frac{l_{max}}{1-\kappa d \sin\left(\frac{\pi}{6}+\phi\right)} \end{bmatrix} = \begin{bmatrix} s^1_{max} \\ s^2_{max} \\ s^3_{max} \end{bmatrix} \qquad (6.22)$$

It can be noticed in (6.22) that the restriction of S is related to l_{max}, κ, and ϕ. The maximum length of extension of the continuum body S_{max} can then be determined by,

$$S_{max}(l_{max}, \kappa, \phi) = \min\left(s^1_{max}, s^2_{max}, s^3_{max}\right) \qquad (6.23)$$

Fig. 6.6 illustrates the maximum length of extension of the continuum body with different parameters and with a constant maximum length of all tendons $l_{max} = 0.3m$. It can be observed that the decrease of S_{max} is mainly due to the increase of the bending curvature κ.

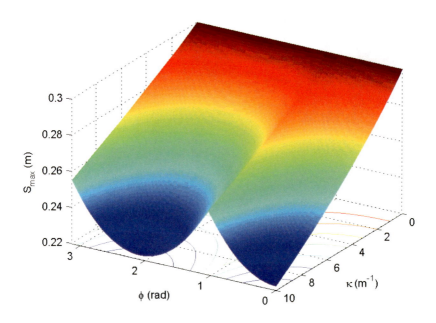

Fig. 6.6 Maximum length of extension of the continuum body with different bending curvature and bending direction at $l_{max} = 0.3m$.

6.2.1.2 Minimum Length of Extension

Let the minimum length of all tendons be l_{min}. According to (6.2), the minimum length of the virtual tendon also equals to l_{min}. However, if we want to achieve certain bending angle θ and bending direction ϕ, it is no longer applicable. To find the restriction of S, (6.1) can be rewritten as,

$$\begin{bmatrix} l_1 \\ l_2 \\ l_3 \end{bmatrix} = \begin{bmatrix} S + \theta d \cos \phi \\ S - \theta d \sin \left(\frac{\pi}{6} - \phi \right) \\ S - \theta d \sin \left(\frac{\pi}{6} + \phi \right) \end{bmatrix} = S + \begin{bmatrix} -\theta d \cos \phi \\ \theta d \sin \left(\frac{\pi}{6} - \phi \right) \\ \theta d \sin \left(\frac{\pi}{6} + \phi \right) \end{bmatrix} \geq \begin{bmatrix} l_{min} \\ l_{min} \\ l_{min} \end{bmatrix} \quad (6.24)$$

6.2 Workspace Analysis

It implies,

$$S \geq \begin{bmatrix} l_{min} + \theta d \cos\phi \\ l_{min} - \theta d \sin\left(\frac{\pi}{6} - \phi\right) \\ l_{min} - \theta d \sin\left(\frac{\pi}{6} + \phi\right) \end{bmatrix} = \begin{bmatrix} s^1_{min} \\ s^2_{min} \\ s^3_{min} \end{bmatrix} \quad (6.25)$$

As a result, the minimum length of extension of the continuum body S_{min} can then be determined by,

$$S_{min}(l_{min}, \theta, \phi) = \max\left(s^1_{min}, s^2_{min}, s^3_{min}\right) \quad (6.26)$$

Fig. 6.7 illustrates the minimum length of extension of the continuum body with different parameters and with a constant minimum length of all tendons $l_{min} = 0m$. It can be observed that the increase of S_{min} is mainly due to the increase of the bending angle θ.

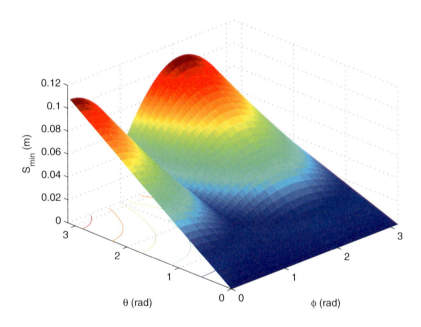

Fig. 6.7 Minimum length of extension of the continuum body with different bending curvature and bending direction at $l_{min} = 0m$.

6.2.1.3 Maximum Bending Curvature

Bending the continuum body requires a certain driving force applied on the tendons to keep the tendons bent. The magnitude of the force relates to the magnitude of the bending curvature and the length of the virtual tendon. Bending the tendons to a greater curvature requires a larger force. In addition, the longer the continuum body, the higher force required to keep the same bending curvature. As a result, the maximum bending curvature is determined by the power of the tendon driving motor and the length of extension of the continuum body.

6.2.1.4 Maximum Climbing Slope

As the additional revolute joint is a passive joint, the maximum climbing slope is determined by the location of the center of mass. Fig. 6.8(a) illustrates the relationship between the location of the center of mass and the limit of the climbing slope. If the climbing slope exceeds the limit, then the rear gripper is pulled out of the gripping substrate by the force of gravity, as illustrated in Fig. 6.8(b). In this case, bending the body can make the rear gripper contact the surface, but there is no means of arranging the rear gripper appress to the tree surface, as shown in Fig. 6.8(c). If the climbing slope does not exceed the limit by too much, then the rear gripper may still be capable of gripping the tree surface and provide sufficient gripping force, as some tolerance of gripping direction is allowed. Hence, Treebot may still be capable of climbing continuously if the front gripper is sufficiently flexible to be appressed to the gripping surface.

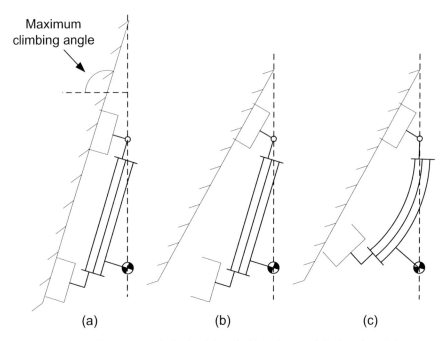

Fig. 6.8 Relationship between the limit of the climbing slope and the location of the center of mass.

6.2 Workspace Analysis

6.2.1.5 Maximal Angle of Twist

As the gripper contains a semi-passive joint, it has certain flexibility to change its direction. In the semi-passive joint, the range of twist about the y- and z-axes are $\pm\pi/3$ and $\pm\pi/6$ respectively. As a result, the workspace of the gripper is a spherical surface, as illustrated in Fig. 6.9. If the gripper can reach a certain position but the required angle of twist exceeds the range, then it is not an admissible climbing position.

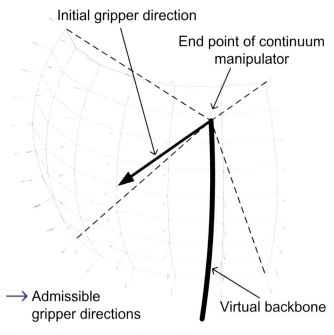

Fig. 6.9 Workspace of the gripper at each reachable position.

6.2.2 Admissible Workspace on a Tree Surface

To determine the motion of Treebot, the admissible climbing workspace on the tree surface must be identified. It is assumed that there is no external force acting on the robot, and its weight is negligible. The compliance effect can thus be neglected. To ensure that the gripper can attach to the tree at a target position, certain physical constraints must not be violated. The maximum inclined angle is a constant constraint. The constraints of the continuum body, including the maximum length of extension and bending curvature, can be determined using the kinematic model. The required angles of twist at a particular position can be obtained from the normal direction of the surface.

In reality, the posture of the continuum manipulator may not equal the analytical result exactly as shown in Fig. 6.5 due to the gravitational force. As mentioned in Chapter 4, the continuum manipulator can be deformed by external force due to the inherent passive compliance. The magnitude of deformation is inversely proportional to the stiffness of the springs and proportional to the weight of Treebot.

6.2.2.1 Required Angle of Twist

By giving a shape a of tree, and the position and orientation of the rear gripper, the angle of twist required to place the front gripper appressed on the tree surface at a target position can be determined. The angle of twist about $^f y$- and $^f z$-axes to appress the front gripper to the target surface can be determined by,

$$\theta_{twist\_y} = \tan^{-1}\left(\frac{^f n_{t\_z}}{^f n_{t\_x}}\right) \tag{6.27}$$

$$\theta_{twist\_z} = \tan^{-1}\left(\frac{^f n_{t\_y}}{^f n_{t\_x}}\right) \tag{6.28}$$

Once appressed, the direction of growth at the new front gripper frame $^f \mathbf{v}_t^a$ can be determined by,

$$^f \mathbf{v}_t^a = \begin{bmatrix} ^f v_{t\_x}^a \\ ^f v_{t\_y}^a \\ ^f v_{t\_z}^a \end{bmatrix} = Rot_z(-\theta_{twist\_z}) Rot_y(-\theta_{twist\_y}) {^f \mathbf{v}_t} \tag{6.29}$$

In addition, the angle between the direction of growth and the z-axis of the new front gripper frame θ_{fx} as illustrated in Fig. 6.10 becomes,

$$\theta_{fx} = \tan^{-1}\left(\frac{^f v_{t\_y}^a}{^f v_{t\_z}^a}\right) \tag{6.30}$$

6.2.2.2 Admissible Target Position

The admissible gripping positions can be determined by considering all of the necessary constraints. Fig. 6.11 to Fig. 6.14 illustrate the admissible gripping positions of the front gripper on a straight tree for different directions of the rear gripper. In the figures, the arrows at the bottom denote the direction of the rear gripper. The inner circle illustrates the circumference of the tree. The dots are the admissible positions of the front gripper with an arrow denoting the direction of the front gripper. This information is useful for determining the motion of Treebot. It can be observed in the figures that when θ_{rx} increases, the admissible angle of change (θ_t) increases and the length of centerline (S_t) decreases accordingly.

6.2 Workspace Analysis

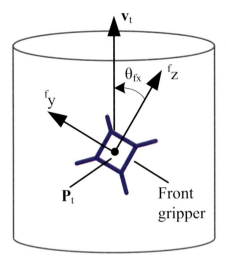

Fig. 6.10 Relationship between the front gripper and the tree model.

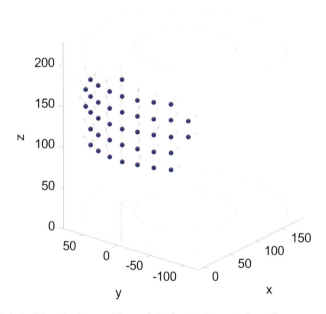

Fig. 6.11 Admissible gripping positions of the front gripper at $\theta_{rx} = 0$.

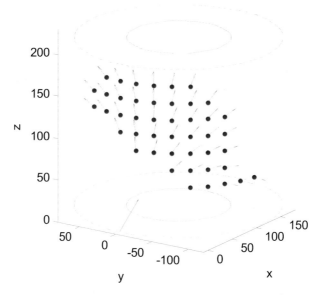

Fig. 6.12 Admissible gripping positions of the front gripper at $\theta_{rx} = \pi/6$.

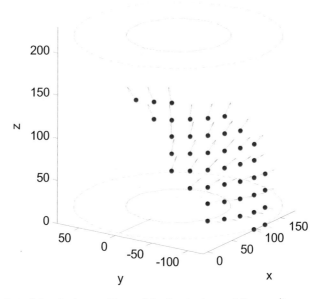

Fig. 6.13 Admissible gripping positions of the front gripper at $\theta_{rx} = \pi/3$.

6.3 Summary

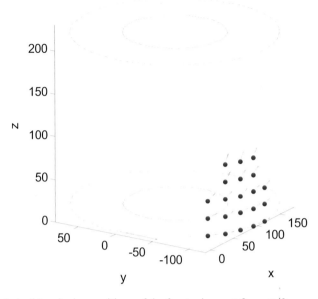

Fig. 6.14 Admissible gripping positions of the front gripper at $\theta_{rx} = \pi/2$.

6.3 Summary

In this chapter, the kinematics of the continuum body was presented, including the relationship among the length of tendons, posture of the continuum body and the Cartesian coordinate at the ends of the continuum body. In addition, the extended kinematics of the continuum body was also proposed. It includes the consideration of the straight sections connected at both ends of the continuum body which is actually the configuration of Treebot.

The workspace analysis presented in this chapter reveals the capability and limitation of Treebot in terms of locomotion. It offers a complete insight to Treebot operators on understanding the admissible climbing motion of Treebot and helps them make decision in manual operation. More importantly, the study of the kinematics and workspace of Treebot are the crucial information for designing path and motion planning algorithms in autonomous control which are going to be discussed in Chapter 7 and 8.

Chapter 7
Autonomous Climbing[1]

The goal of developing of Treebot is to assist or replace people in performing forestry tasks on trees. A certain level of autonomous climbing ability of Treebot helps reduce the complexity of operation by users. An autonomous climbing strategy for Treebot is thus proposed. To determine the motions to climb up autonomously in an unknown environment, a robot must be equipped with sensors that can explore the environment. Vision sensors provide rich information about the environment. However, processing the data require a great deal of computational power. Moreover, light conditions vary in outdoor environments, which will affect the accuracy of visual information. There are many living creatures that do not rely on visual information, but can navigate well in their living environment. Inchworms, for example, navigate on trees by using their sense of touch only. Although the information obtained by tactile sensors is not rich, it is reliable. Furthermore, the processing of tactile information is much simpler than that of visual information. As a result, inspired by arboreal animals, an algorithm is developed to allow Treebot to climb irregularly shaped trees autonomously by using tactile sensors and a tilting sensor only. The development of the algorithm can reveal how tactile sensors can best be employed in autonomous tree climbing.

This chapter is organized as follows. Section 7.1 introduces the structure of the proposed autonomous climbing strategy. Section 7.2 proposes a tree shape approximation method. In Section 7.3, the motion planning strategy is discussed. The experimental results are presented in Section 7.4. Finally, a summary is provided in Section 7.5.

---

[1] Portions reprinted, with permission from Tin Lun Lam, and Yangsheng Xu, "Climbing Strategy for a Flexible Tree Climbing Robot - Treebot", IEEE Transactions on Robotics. ©[2011] IEEE.

7.1 Autonomous Climbing Strategy

Robots may topple sideways when climbing on an inclined tree. The optimal climbing position to avoid this tendency is above the centerline of the tree, so that the gravitational force acts on the robot to direct it to the centerline of tree [58]. In the following text, "upper apex" is used to describe this optimal position. The autonomous climbing algorithm aims to make Treebot climb a tree along an optimal path. The procedure for the autonomous climbing motion is shown in Fig. 7.1. It mainly includes the works of exploration, tree shape approximation, and motion planning. It is assumed that Treebot is already attached to a tree by the rear gripper, that the front gripper is detached, and that the continuum body is contracted to the minimum length. Completing the main loop of the procedure once is termed as a complete climbing gait. By repeating the climbing gait, Treebot can climb a tree along the optimal path. The following sections discuss this procedure in detail.

7.2 Tree Shape Approximation

The concept of tree shape modeling is mentioned in Chapter 6. This section discusses the method used to approximate the values of the model parameters. Treebot explores the shape of a tree by tentacles (which form by tactile sensors) attached to the front gripper, and uses the exploration data to approximate the shape of the tree. The exploring motion of Treebot is based on the proposed exploring strategy. Approximating the shape of the explored portion of the tree is useful to determine the location of the optimal climbing position and to predict the shape of the tree ahead for motion planning. There are many techniques for shape reconstruction using information from tactile sensors. Okamura and Cutkosky [56] proposed a method for extracting the local features of a surface. Jia and Tian [55] reconstructed the unknown local curved surface by using one-dimensional tactile data. Schopfer [57] used a 2D pressure array to reconstruct the unknown shape of an object. All of these methods can successfully reconstruct the unknown shape of an object. Here, as the geometry of a branch is assumed to be a curved cylinder, an efficient reconstruction scheme can be developed based on a known geometric model to speed up the exploration and reconstruction processes.

7.2.1 Exploring Strategy

The proposed exploring strategy aims to trace a growth path of a tree using the front gripper, which is similar to the feature-tracing method presented in [56]. The trajectory of the front gripper can then be used to reconstruct the shape of the tree. The top left and right tentacles attached to the front gripper are used for exploring. The state and action pairs for the exploring motion are listed in Table 7.1. The forward

7.2 Tree Shape Approximation

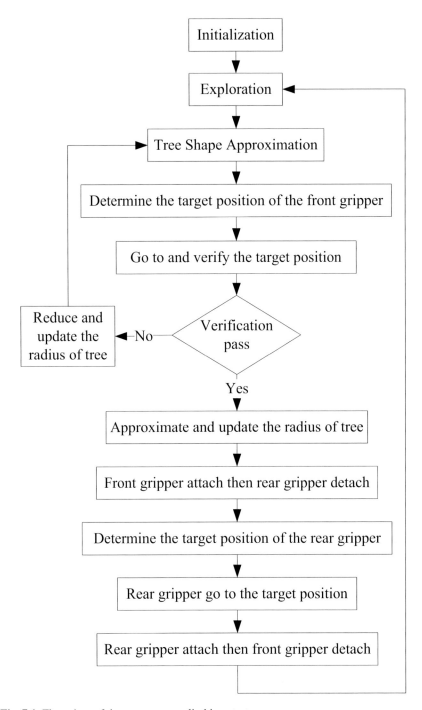

Fig. 7.1 Flow chart of the autonomous climbing strategy.

and left directions are defined as the positive x^r and y^r directions, respectively. A tentacle acts in a similar fashion to a mechanical switch. It is triggered when a force acts on the bottom part of the tentacle over a certain threshold.

Table 7.1 Exploring strategy

State	Action
No tentacle is triggered	Bend forward with extension
Both tentacles are triggered	Bend backward with extension
Only left tentacle is triggered	Bend left and backward
Only right tentacle is triggered	Bend right and backward
Length of extension reaches a constant value	Finish exploration

In the exploring strategy, the front gripper approaches and leaves the tree surface repeatedly. When the front gripper leaves the growth path of a tree, only one side of the tentacle is triggered frequently. The front gripper then moves to eliminate this unbalanced triggering between tentacles so as to keep the front gripper follow the growth path of the tree.

Once a tentacle is triggered, the Cartesian coordinates of the front gripper are recorded, which can be found by (6.5). During the exploring motion, the semi-passive joint is locked to make (6.5) applicable. As only one tactile sensor is installed on each tentacle, there is no way of determining where a force is exactly applied along the tentacles. As a result, the triggering of a tentacle does not necessarily indicate that the center of the front gripper is placed on the tree surface. To obtain accurate data points, the selected points must include only those points at which both the left and right tentacles are triggered at the same time, or the average position of the points at which the left and right tactile sensors are triggered alternatively.

7.2.2 Arc Fitting

As the shape of a tree is approximated as a perfect cylinder with an uniform bend, the data acquired from exploration are fitted with a 3D arc to help reconstruct the shape of the tree. It is set that the arc crosses the first and last data points. As a result, the data are transformed such that the first data point is on the origin and the last data point is on the z-axis (rotation about the z-axis of $-\theta_z$ and then rotate about the y-axis of $-\theta_y$) as shown in Fig. 7.2(b).

7.2 Tree Shape Approximation

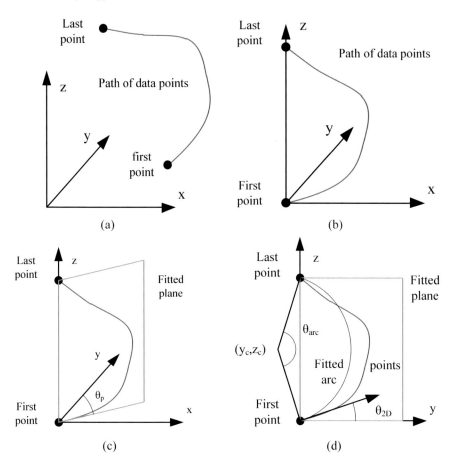

Fig. 7.2 Notations and procedures for arc fitting. (a) Path segment; (b) Transformation; (c) Plane fitting; (d) Arc fitting.

7.2.2.1 Plane Fitting

To simplify the 3D arc fitting problem into 2D, the data are fitted onto a plane, as illustrated in Fig. 7.2(c). This is accomplished by determining the optimal angle of rotation about the z-axis, (θ_p) to minimize the absolute x component value of the transformed data. The path is then rotate θ_p about the z-axis and projected on the $y-z$ plane thus becomes a 2D path, as illustrated in Fig. 8.5(d). Let $[x_i, y_i, z_i]$ be the transformed points where $i \in [1, \eta]$ and η is the number of data points, θ_p can be obtained by minimizing:

$$x_i \cos \theta_p - y_i \sin \theta_p \tag{7.1}$$

By using the least square method and considering all of the points:

$$\sum \frac{d}{d\theta_p}(x_i\cos\theta_p - y_i\sin\theta_p)^2 = 0$$

$$-2\sum(x_i\cos\theta_p - y_i\sin\theta_p)(x_i\sin\theta_p + y_i\cos\theta_p) = 0$$

$$(\cos^2\theta_p - \sin^2\theta_p)\sum(x_iy_i) = \sin\theta_p\cos\theta_p\sum(y_i^2 - x_i^2)$$

$$\frac{\cos^2\theta_p - \sin^2\theta_p}{\sin\theta_p\cos\theta_p} = \frac{\sum(y_i^2 - x_i^2)}{\sum(x_iy_i)}$$

$$\frac{\cos\theta_p}{\sin\theta_p} - \frac{\sin\theta_p}{\cos\theta_p} = \frac{\sum(y_i^2 - x_i^2)}{\sum(x_iy_i)}$$

$$\frac{1}{\tan\theta_p} - \tan\theta_p = \frac{\sum(y_i^2 - x_i^2)}{\sum(x_iy_i)}$$

$$\tan^2\theta_p + \frac{\sum(y_i^2 - x_i^2)}{\sum(x_iy_i)}\tan\theta_p - 1 = 0$$

It implies,

$$\theta_p = \tan^{-1}\left(\frac{-o \pm \sqrt{o^2 + 4}}{2}\right) \quad (7.2)$$

where $o = \frac{\sum y_i^2 - \sum x_i^2}{\sum x_iy_i}$.

In addition, the fitness value of the plane fitting is defined as:

$$e_{plane} = \sqrt{\frac{\sum(x_i\cos\theta_p - y_i\sin\theta_p)^2}{\eta}} \quad (7.3)$$

The lower the e_{plane} indicates the better plane fitting result.

7.2.2.2 2D Arc Fitting

Once the fitted plane has been obtained, the data are projected onto the plane and rotate about z-axis onto the y-z plane. Then, referring to Fig. 7.2(d), the center of the approximated arc (y_c, z_c), the curvature of the bend κ_{arc}, and the angle of the arc θ_{arc} can be found by using a 2D arc fitting method. (y_c, z_c) and r be the center and radius of the approximated arc respectively. In the arc fitting, it is assumed that the approximated arc must pass through two selected points (y_a, z_a) and (y_b, z_b) in the data points.

To cross the selected points, the approximated arc should fulfill the following equations:

$$(y_a - y_c)^2 + (z_a - z_c)^2 = r^2 \quad (7.4)$$

$$(y_b - y_c)^2 + (z_b - z_c)^2 = r^2 \quad (7.5)$$

7.2 Tree Shape Approximation

Combining (7.4) and (7.5):

$$(y_a - y_c)^2 + (z_a - z_c)^2 = (y_b - y_c)^2 + (z_b - z_c)^2$$

$$2(z_b - z_a)z_c = (y_b^2 + z_b^2) - (y_a^2 + z_a^2) + 2(y_a - y_b)y_c$$

$$z_c = \frac{(y_b^2 + z_b^2) - (y_a^2 + z_a^2)}{2(z_b - z_a)} + \frac{(y_a - y_b)}{(z_b - z_a)} y_c$$

Let,

$$z_c = a + b y_c \qquad (7.6)$$

where $a = \frac{(y_b^2 + z_b^2) - (y_a^2 + z_a^2)}{2(z_b - z_a)}$ and $b = \frac{y_a - y_b}{z_b - z_a}$.

The distance error e_i of a point to the approximated arc can be found by:

$$e_i = (y_i - y_c)^2 + (z_i - z_c)^2 - r^2 \qquad (7.7)$$

Sub. (7.4) into (7.6):

$$e_i = (y_i - y_c)^2 + (z_i - z_c)^2 - \left[(y_a - y_c)^2 + (z_a - z_c)^2\right]$$
$$= (y_i^2 + z_i^2) - (y_a^2 + z_a^2) + 2(y_a - y_i)y_c + 2(z_a - z_i)z_c \qquad (7.8)$$

Sub. (7.6) into (7.8):

$$e_i = (y_i^2 + z_i^2) - (y_1^2 + z_1^2) + 2(y_1 - y_i)y_c + 2(z_1 - z_i)(a + b y_c)$$
$$= (y_i^2 + z_i^2) - (y_1^2 + z_1^2) + 2a(z_1 - z_i) + 2[(y_1 - y_i) + b(z_1 - z_i)]y_c \qquad (7.9)$$

Let,

$$e_i = m_i + n_i y_c \qquad (7.10)$$

where $m_i = (y_i^2 + z_i^2) - (y_a^2 + z_a^2) + 2a(z_a - z_i)$ and $n_i = 2[(y_a - y_i) + b(z_a - z_i)]$.
By using the least square method and considering all of the points, that is,

$$\frac{d}{dy_c} \sum e_i^2 = \frac{d}{dy_c} \sum (m_i + n_i y_c)^2 = 0 \qquad (7.11)$$

It implies,

$$\sum (m_i + n_i y_c) n_i = 0$$
$$\sum m_i n_i + \sum n_i^2 y_c = 0$$
$$y_c = -\frac{\sum m_i n_i}{\sum n_i^2} \qquad (7.12)$$

Sub. (7.12) into (7.6):

$$z_c = a - b \frac{\sum m_i n_i}{\sum n_i^2} \qquad (7.13)$$

Sub. (7.12) and (7.13) into (7.4):

$$r = \sqrt{\left(y_a + \frac{\sum m_i n_i}{\sum n_i^2}\right)^2 + \left(z_a - a + b\frac{\sum m_i n_i}{\sum n_i^2}\right)^2} \qquad (7.14)$$

The fitness of the fitted arc is measured by:

$$e_{arc} = \sqrt{\frac{1}{\eta}\sum e_i^2} = \sqrt{\frac{1}{\eta}\sum \left(m_i + n_i\frac{\sum m_i n_i}{\sum n_i^2}\right)^2} \qquad (7.15)$$

The lower the e_{arc} indicates the better arc fitting result. Additionally, the parameters of the fitted arc can be obtained by:

$$\kappa_{arc} = \frac{1}{r} \qquad (7.16)$$

$$\theta_{arc} = 2\cos^{-1}(y_c \kappa_{arc}) \qquad (7.17)$$

The two selected points (y_a, z_a) and (y_b, z_b) are defined as the first and last data points, i.e., $(0,0)$ and (y_η, z_η) respectively.

Finally, the tangent vector $\mathbf{v}_S$ and the bending direction $\mathbf{v}_{bend}$ (toward the center of the bend) of the arc at the starting point in the rear gripper frame can be determined by:

$$\mathbf{v}_S = Rot_z(\theta_z)Rot_y(\theta_y)Rot_z(-\theta_p)\begin{bmatrix}0\\\cos\theta_{2D}\\\sin\theta_{2D}\end{bmatrix} \qquad (7.18)$$

$$\mathbf{v}_{bend} = Rot_z(\theta_z)Rot_y(\theta_y)Rot_z(-\theta_p)\begin{bmatrix}0\\-\sin\theta_{2D}\\\cos\theta_{2D}\end{bmatrix} \qquad (7.19)$$

where $\theta_{2D} = \tan^{-1}\frac{z_c}{y_c} + sign(y_c)\frac{\pi}{2}$.

7.2.3 Tree Shape Reconstruction

To approximate the parameters of the tree model, the fitted arc is transformed into the tree frame, that is, to transform the tangent vector of the arc on the $^T z$-axis:

$$^T\mathbf{v}_S = \begin{bmatrix}^T v_{S\_x}\\^T v_{S\_y}\\^T v_{S\_z}\end{bmatrix} = Rot_x(\theta_{rx})Rot_y(\theta_{ry})\mathbf{v}_S \qquad (7.20)$$

$$^T\mathbf{v}_{bend} = \begin{bmatrix}^T v_{bend\_x}\\^T v_{bend\_y}\\^T v_{bend\_z}\end{bmatrix} = Rot_x(\theta_{rx})Rot_y(\theta_{ry})\mathbf{v}_{bend} \qquad (7.21)$$

where $\theta_{ry} = \sin^{-1}(^T v_{S\_x})$ and $\theta_{rx} = -\tan^{-1}\frac{^T v_{S\_y}}{^T v_{S\_z}}$.

7.2 Tree Shape Approximation

In addition, the bending direction of the fitted arc in the tree frame is,

$$\phi_{arc} = \tan^{-1} \frac{^T v_{bend\_y}}{^T v_{bend\_x}} \tag{7.22}$$

According to Fig. 7.3, by giving the radius of a tree R_{tree}, the values of the parameters of the tree model, i.e., ϕ_{tree}, κ_{tree} and S_{tree} can be determined as:

$$\phi_{tree} = \phi_{arc} \tag{7.23}$$

$$\kappa_{tree} = \frac{1}{\left(\frac{1}{\kappa_{arc}} - (h_g + R_{tree})\cos\phi_{tree}\right)} \tag{7.24}$$

$$S_{tree} = \frac{\theta_{arc}}{\kappa_{tree}} \tag{7.25}$$

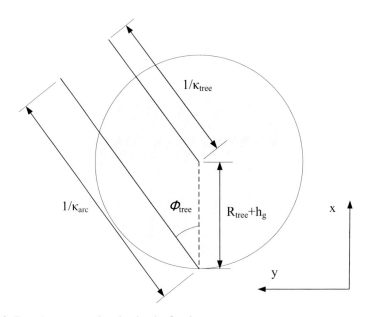

Fig. 7.3 Tree shape approximation by the fitted arc.

7.2.4 Angle of Change to the Upper Apex

To find the upper apex of a tree, the direction of gravity must first be established. The gravity vector can be obtained from the tilting sensor attached to the front gripper. As the tilting sensor is fixed to the front gripper, the coordinates with respect to the rear gripper frame can be determined by the posture of Treebot, as discussed in Chapter 6. Once the tree shape has been approximated, the transformation relationship between the rear gripper frame and the tree frame (θ_{rx} and θ_{ry}) can be obtained. The gravity vector can then be represented in the target position frame ($^tx - {^ty} - {^tz}$) as illustrated in Fig. 7.4. The angle of change required to reach the upper apex $\theta_{optimal}$ is equivalent to the angle of rotation about ^tz-axis required to make the gravity vector $\mathbf{v}_{gravity}$ lies on the $^tx - {^tz}$ plane with a positive x component, i.e.,

$$\theta_{optimal} = \tan^{-1}\left(\frac{v_{gravity\_y}}{v_{gravity\_x}}\right) \tag{7.26}$$

where $\mathbf{v}_{gravity} = \begin{bmatrix} v_{gravity\_x} & v_{gravity\_y} & v_{gravity\_z} \end{bmatrix}$.

In addition, the inclined angle of the tree can be obtained by:

$$\varphi_{incline} = \sin^{-1}\left(|v_{gravity\_z}|\right) \in \left[0, \frac{\pi}{2}\right] \tag{7.27}$$

7.2.5 Tree Radius Approximation

The data from the exploring motion can be used to approximate the shape of tree but not the radius of tree. Two methods are thus proposed to approximate the radius.

7.2.5.1 Method 1

This method approximates the radius of a tree by comparing the angle of change to the upper apex ($\theta_{optimal}$) which are obtained by two different positions on the tree surface. As obtaining $\theta_{optimal}$ does not require the radius of the tree to be known, the new approximated tree radius R'_{tree} can be obtained as follow:

$$R'_{tree} = R_{tree}\left|\frac{\hat{\theta}}{\Delta\theta_{optimal}}\right| \tag{7.28}$$

where $\Delta\theta_{optimal}$ is the difference between $\theta_{optimal}$ obtained by two different positions on the tree surface. R_{tree} is the last approximated radius of the tree. $\hat{\theta}$ is the angle of change to the second position. This method updates the information of the

7.2 Tree Shape Approximation

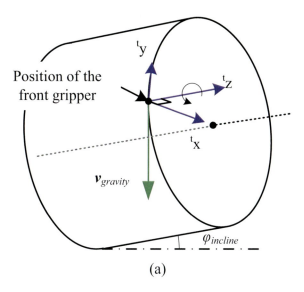

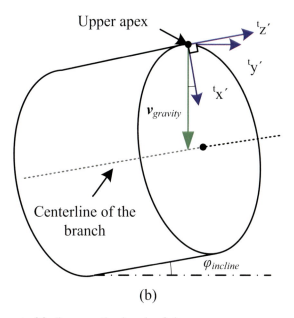

Fig. 7.4 The concept of finding an optimal angle of change.

radius of the tree at every climbing gait. However, the application of this method is not feasible when the inclined angle of the tree is $\pi/2$. It is because in this state the slope in any position is minimal and finding $\theta_{optimal}$ will result in a trivial solution. In such a case, the tree radius can be approximated using Method 2.

7.2.5.2 Method 2

This method is based on the unsuccessful placement of the front gripper in the target position, as this indicates that the actual radius of the tree must be smaller than the approximated radius. Once the front gripper fails to appress to the target position, the approximated radius of the tree is then reduced by a certain value, and the maximum angle of change is recalculated. The target position is updated according to the adjusted maximum angle of change and then the gripper goes to the updated target position again. This trial process repeats until the front gripper appresses to the updated target position successfully.

7.3 Motion Planning

The optimal solution for making Treebot climb on the upper apex of a tree is to place the front gripper on and at the same time set the direction of the gripper parallel to the path of the upper apex. The rear gripper can then be placed on the upper apex by a contraction motion. As Treebot is a nonholonomic system [43] that the direction of the gripper and the position of the gripper are coupled, it is always impossible to achieve a specific position and direction at the same time in one climbing gait. As a result, three feasible motion planning strategies are proposed to achieve the goal as follows. Each of the motion planning strategy has its own merits and drawbacks.

7.3.1 Strategy 1

This strategy uses two climbing gaits to put the front gripper on the target position and direction. In the first climbing gait, the direction of the rear gripper is adjusted, such that in the second climbing gait, the front gripper can set on the target position and direction. Fig. 7.5 illustrates the concept of achieving the target position and direction in two climbing gaits. In the figures, the circles denote the target position and the arrow represents the target direction. Rectangles colored in white and grey represents the attached and detached grippers respectively. After an exploration (a), Treebot acquires the optimal position and direction for the front gripper. The front gripper returns to the original position by fully contract the continuum body (b), then the direction of the rear gripper adjusts (c). Finally, the front gripper moves to the target position in the appropriate position and direction (d). The forward motion

7.3 Motion Planning

is completed when the continuum body contracts to pull up the rear gripper (e). It can be seen that although this strategy can achieve the target position and direction exactly, it takes two extend-contract motions to move forward, which is quite time-consuming. In addition, the rear gripper must first move farer to the target position as shown in Fig. 7.5(c), it increases the tendency of toppling sideways.

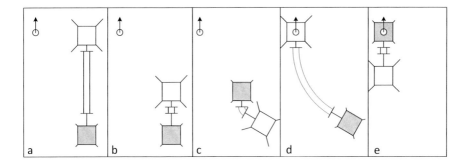

Fig. 7.5 Series of motions by Strategy 1.

7.3.1.1 Motion for the Front Gripper

As has been stated, the optimal solution to ensure that Treebot follows the best path is to place the front gripper directly on the upper apex. However, it is necessary to consider that when the inclined angle is large (nearly vertical), a change of position will not reduce much of the pull-out force generated by the gravitational force. As a result, to avoid Treebot having to make a large change in angle to reduce the pull-out force by only a small amount, the magnitude of the angle of change should decrease when the inclined angle of the tree is large. As a result, the target angle of change is defined as:

$$\theta_t = \theta_{optimal}\left[1 - \frac{\varphi_{incline}}{\pi/2}\right] \quad (7.29)$$

If θ_t exceeds the admissible angle of change, then it is replaced by the admissible angle of change that is closest to θ_t. With θ_t and let $S_t = S_{tree}$, the target position in view of tree frame $^T\mathbf{P}_t$ can be determined by (6.13). The position of the rear gripper in the target position frame ($^tx - {^ty} - {^tz}$) is then obtained as,

$$^t\mathbf{P}_r = -{^r\mathbf{P}_t} \quad (7.30)$$

It is assumed that the rear gripper is already at the position as illustrated in Fig. 7.5(d). As the target position frame and the front gripper frame is identical when the front gripper is appressed to the target position, the posture of the continuum body to place the front gripper to the target position and direction $\left(S_f, \kappa_f, \phi_f\right)$ can be determined by (6.8) and setting $\left[{}^f x_r, {}^f y_r, {}^f z_r\right] = {}^t\mathbf{P}_r$.

7.3.1.2 Motion for the Rear Gripper

According to (6.9), the target direction of the rear gripper at the target position frame ${}^t\mathbf{v}_{tr}$ is defined as,

$$ {}^t\mathbf{v}_{tr} = Rot_z\left(\phi_f\right) Rot_y\left(-\kappa_f S_f\right) Rot_z\left(-\phi_f\right) \begin{bmatrix} 0 \\ 0 \\ 1 \end{bmatrix} \quad (7.31)$$

Based on (6.14), the target direction of the rear gripper at the tree frame ${}^T\mathbf{v}_{tr}$ can be obtained by,

$$ {}^T\mathbf{v}_{tr} = \begin{bmatrix} {}^T v_{tr\_x} \\ {}^T v_{tr\_y} \\ {}^T v_{tr\_z} \end{bmatrix} = Rot_z\left(\phi_{tree}\right) Rot_y\left(\kappa_{tree} S_{tree}\right) Rot_z\left(-\phi_{tree}\right) Rot_z\left(\theta_t\right) {}^t\mathbf{v}_{tr} \quad (7.32)$$

It is assumed that the direction of the front gripper is same as the rear gripper at the state as shown in Fig. 7.5(b). The posture of the continuum body to place the rear gripper (S_r, κ_r, ϕ_r) to the target position as illustrated in Fig. 7.5(c) can then be obtained by,

$$\phi_r = -sign\left(\theta_{tr} - \theta_{rx}\right) \frac{\pi}{2} \quad (7.33)$$

$$\kappa_r = \frac{|\theta_{tr} - \theta_{rx}|}{S_r} \quad (7.34)$$

$$S_r = S_{min} \quad (7.35)$$

In that,

$$\theta_{tr} = \tan^{-1}\left(\frac{{}^T v_{tr\_y}}{{}^T v_{tr\_z}}\right) \quad (7.36)$$

S_{min} is the minimum length of extension of the continuum body to achieve the specified bending angle defined by (6.26).

7.3.2 Strategy 2

In view of the slow speed and the increase of the tendency of toppling sideways in some motions by Strategy 1, Strategy 2 is proposed as illustrated in Fig. 7.6 to eliminate those drawbacks. In the figures, the dotted circle and arrow denote the next target position and the target direction respectively. As shows in the figures, after

7.3 Motion Planning

exploration (a), the front gripper moves directly to the target position by neglecting the target direction (b). The continuum body then contracts and adjusts the direction of the rear gripper to make it parallel to the direction of growth of the tree (c). The parallel placing of the rear gripper is aimed to make the exploring posture of Treebot similar to the shape of tree such that Treebot can explore farer by each exploration motion. It also avoids the rear gripper griper farer to the target position to eliminate the tendency of toppling sideways. However, this motion cannot make the front gripper go to but just closer to the target position and direction then before. By implementing this motion planning strategy repeatedly (d) (e), Treebot will get closer and closer to the upper apex and at the same time the direction of Treebot is parallel to the growing direction of the tree. This motion planning strategy moves forward by one extend-contract motion only which is two times faster than Strategy 1. The drawback of Strategy 2 is that it takes time to converge to the upper apex.

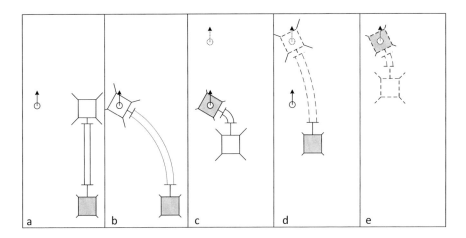

Fig. 7.6 Series of motions by Strategy 2.

7.3.2.1 Motion for the Front Gripper

The target position of the front gripper in this strategy is same as that in Strategy 1. In view of the rear gripper frame, the target position $^r\mathbf{P}_t$ can be determined by (6.18). Once the target position $^r\mathbf{P}_t$ is defined, the posture of the continuum body (S_f, κ_f, ϕ_f) to place the front gripper can be obtained by (6.6).

7.3.2.2 Motion for the Rear Gripper

To place the rear gripper parallel to the direction of growth of the tree, the angle between the direction of the front gripper and the growth direction of tree θ_{fx} after appressed as illustrated in Fig. 7.6(c) should be determined and it can be found by

(6.30). Then, the posture of the continuum body (S_r, κ_r, ϕ_r) to set the rear gripper in the appropriate direction from the target position can be determined as,

$$\phi_r = -\text{sign}(\theta_{fx}) \frac{\pi}{2} \tag{7.37}$$

$$\kappa_r = \frac{|\theta_{fx}|}{S_r} \tag{7.38}$$

$$S_r = S_{min} \tag{7.39}$$

S_{min} is defined by (6.26) which is the minimum length of extension of the continuum body to achieve the specified bending angle.

7.3.3 Strategy 3

In view of the slow convergence of Strategy 2, this motion planning strategy is proposed. This strategy is similar to Strategy 2 except the angle adjustment of the rear gripper is different. The motions of Strategy 3 are illustrated in Fig. 7.7. After exploration (a), the front gripper moves directly to the target position by neglecting the target direction (b). The continuum body then contracts and adjusts the direction of the rear gripper such that the front gripper can move to the next target position and direction (marked as a dotted circle and arrow respectively) in the next climbing gait (c), (d), (e). The next target position and direction are approximated from the current information. This strategy requires two extend-contract motions to place the front gripper in the future target position and direction. It moves two times faster than Strategy 1 and similar to Strategy 2. However, the drawback of this method is that it may not go exactly to the target position and direction due to inaccurate estimation of the future target position and direction.

7.3.3.1 Motion for the Front Gripper

The motion for the front gripper in Strategy 3 is same as that in Strategy 2.

7.3.3.2 Motion for the Rear Gripper

To determine the target position of the rear gripper, it is necessary to approximate the future target position and direction of the front gripper. The future length from the current rear gripper position S'_{tree} is approximated as,

$$S'_{tree} = S_{tree} + S_{explore} \tag{7.40}$$

7.3 Motion Planning

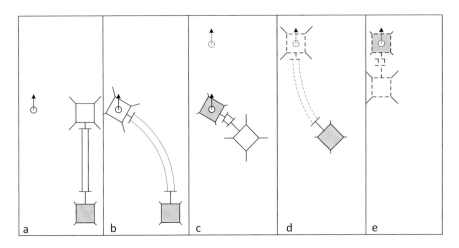

Fig. 7.7 Series of motions by Strategy 3.

where S_{tree} is the length of the current approximated segment of tree. $S_{explore}$ is the approximated future length to be explored which is defined as:

$$S_{explore} = S_{tree} - l_f - l_r \qquad (7.41)$$

The future optimal position can then be obtained by the same method of finding the target optimal position for the placement of the front gripper.

Once the future target position and direction of the front gripper have been obtained, it is necessary to find the position of the continuum body that the front gripper is placed on the current target position. The position in the tree frame can be determined by:

$$^T\mathbf{p}_b = {^T}\mathbf{p}_t - l_f{^T}\mathbf{v}_f \qquad (7.42)$$

where $^T\mathbf{v}_f$ is the direction of the front gripper in the current target position.

To determine the posture of the continuum body from the future target position and direction of the front gripper to place the rear gripper to the target position of the continuum body, it is first transform $^T\mathbf{p}_b$ to the future front gripper frame $^{ff}\mathbf{p}_b$, then find the posture of the continuum body (θ_f and ϕ_f) to place its rear part to the $^{ff}\mathbf{p}_b$ by (6.8).

The direction of the rear gripper in the future front gripper frame $^{ff}\mathbf{v}_r$ can be found by:

$$^{ff}\mathbf{v}_r = Rot_z(\phi_f) Rot_y(-\theta_f) Rot_z(-\phi_f) \begin{bmatrix} 0 \\ 0 \\ 1 \end{bmatrix} \qquad (7.43)$$

The angle between the direction of the rear gripper and the growth direction of the tree in the target position γ_r can then be determined by transforming $^{ff}\mathbf{v}_r$ to the coordinate frame of the target position. To make the exploring motion easy to implement, γ_r should be as small as possible. As a result γ_r is bounded in $\pm\pi/4$.

Finally, the posture of the continuum body to place the rear gripper in the appropriate direction from the target position (S_r, κ_r, ϕ_r) is determined by:

$$\kappa_r = \frac{|\gamma_f - \gamma_r|}{S_r} \qquad (7.44)$$

$$\phi_r = \frac{\gamma_f - \gamma_r}{|\gamma_f - \gamma_r|} \frac{\pi}{2} \qquad (7.45)$$

$$S_r = S_{min} \qquad (7.46)$$

where γ_f denotes the angle between the direction of the front gripper and the direction of growth of the tree in the target position. S_{min} is defined by (6.26) which is the minimum length of extension of the continuum body to achieve the specified bending angle.

7.3.4 Verification of Target Position

The gripper may not be able to appress to the target position, which may result in an inaccurate approximation or change in the radius of the tree. The signals from tentacles can be used to detect whether the gripper is appressed to the surface of tree. The gripper is regarded as appressed when any two tentacles are triggered diagonally.

When the gripper is on the target position, Treebot attempts to appress the gripper to the tree surface. The semi-passive joint is first unlocked so that the gripper can be rotated freely. The gripper then pushes forward into the tree surface a certain distance to try to appress the gripper to the tree surface. If it cannot be appressed to the surface, the target position is inadmissible. In this case, the approximated radius of the tree is reduced and the target position is recalculated. The process is then repeated until an appressed placement is achieved.

7.4 Experiments

Numerous experiments have been carried out to evaluate the proposed autonomous climbing algorithm in terms of tree shape approximation, optimal path following, and climbing a tree with branches.

7.4 Experiments

7.4.1 Tree Shape Approximation

The results of three experiments to test tree shape approximation, Test 1, Test 2 and Test 3, are shown in Fig. 7.8, 7.9 and 7.10 respectively. The subfigures (a) show the approximation target and the final exploring posture of Treebot, and the subfigures (b) illustrate the approximation result. In the subfigures (b), the solid arc denotes the posture of Treebot, the dots are the exploration data, the dotted arc represents the fitted arc and the large circles represent the circumference of the ends of the segment of tree. The parameters for the approximated shape of the tree are listed at the top of the figures. The dashed line in the subfigures (a) denotes the centerline of the tree.

In Test 1, the explored tree segment is straight, and the direction of the rear gripper is parallel to the direction of growth of the tree. In the approximation result, the shape of the tree is almost straight which approximate the actual shape of tree correctly.

The setting of Test 2 is the same as that of Test 1, except that the direction of the rear gripper is not parallel to the direction of growth of the tree. It can be seen that

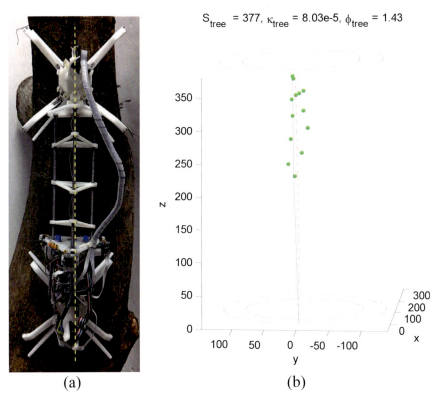

Fig. 7.8 Test 1: Tree shape approximation on a straight tree. (a) Approximation target and final exploring posture of Treebot; (b) Approximation result.

although the posture of Treebot does not match the shape of the tree, the tree shape can still be approximated correctly. It shows that the approximation is independent to the direction of the rear gripper. This property is important to make the tree shape approximation feasible in arbitrary posture of Treebot.

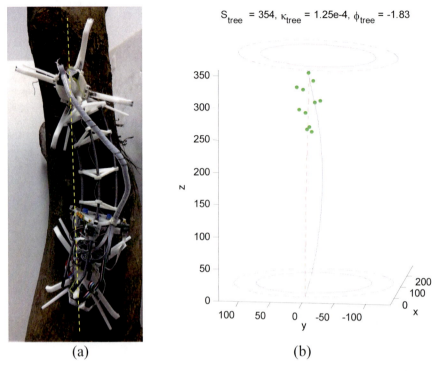

Fig. 7.9 Test 2: Tree shape approximation on a straight tree. (a) Approximation target and final exploring posture of Treebot; (b) Approximation result.

In Test 3, the explored tree segment is bent leftward and backward. It can be observed that the approximated shape of the tree is also bent in a similar fashion to the actual shape of the tree (the approximated bending direction is 2.24rad). This demonstrates that the algorithm can approximate a bent tree in 3D correctly.

7.4.2 Motion Planning

The proposed motion planning strategies guide Treebot to climb on an upper apex of trees. Since Strategy 2 and Strategy 3 are more practical as they have faster climbing speed, experiments were conducted using Strategy 2 and Strategy 3 to evaluate their performance.

7.4 Experiments

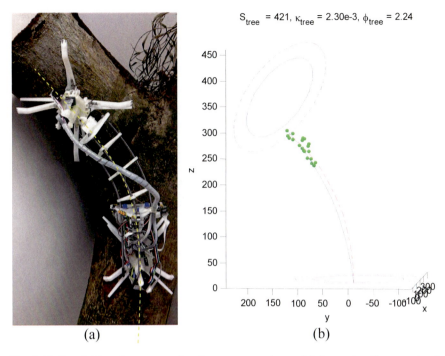

Fig. 7.10 Test 3: Tree shape approximation on a curved tree. (a) Approximation target and final exploring posture of Treebot; (b) Approximation result.

The first experiment was conducted on a tree with a 70-degree inclined angle to evaluate the performance of Strategy 2. Fig. 7.11 shows some of the corresponding climbing motions. The dashed lines denote the upper apex of the tree. Initially, Treebot was not on the upper apex but parallel to the direction of growth of the tree. After one climbing gait, the whole body of Treebot got closer a bit to the upper apex as illustrated in Fig. 7.11(c). It totally took five climbing gaits to climb on the upper apex. The motions were similar to those proposed and illustrated in Fig. 7.6. This result indicates that Strategy 2 can successfully guide Treebot to climb along the upper apex of trees. It can be observed that the climb-up distances were shorter at the beginning of the climbing gaits, while the climb-up distance was obviously longer at the last climbing gait as illustrated in Fig. 7.11(g) and (h). It is because at the beginnings, the angle between the direction of the front gripper and the growth direction of tree θ_{fx} was large, most of the extension of the continuum body were used for adjusting the position but not much for climbing up. At the later climbing gaits, θ_{fx} became smaller as Treebot got close to the upper apex by the previous movements. Hence, larger portions of the extension motions could be used for climbing up.

Fig. 7.11 Experimental result of going to optimal path by Strategy 2.

7.5 Summary

Another experiment was conducted to evaluate the performance of Strategy 3 in which Treebot was commanded to climb the same tree with identical settings. Fig. 7.12 shows the corresponding exploring and climbing motions. The upper apex is marked as dashed line in the figures. Initially, Treebot was not on the optimal path but parallel to the direction of growth of the tree. Different to Strategy 2, the orientation of the rear gripper was not parallel to the direction of grow of the tree in the climbing motions but had a certain angle as seen in Fig. 7.12(d). The specific orientation of the rear gripper permits the front gripper placed on the upper apex and at the same time the direction of the front gripper parallel to the direction of growth of the tree in the next step as seen in Fig. 7.12(f). The rear gripper then aligned on the upper apex too by fully contracting the continuum body as shown in Fig. 7.12(g). The experimental result shows that after two climbing gaits, the whole body of Treebot was successfully placed on the optimal path with motions similar to those proposed in the motion planning strategy illustrated in Fig. 7.7. It also shows that Strategy 3 is faster than Strategy 2 in guiding Treebot climbing on the upper apex.

7.4.3 Climbing a Tree with Branches

An experiment was conducted to evaluate the motion of Treebot on a tree with branches. In the first test, the initial position of Treebot is shown in Fig. 7.13(a). Treebot selected branch A to climb, as shown in Fig. 7.13(b). In the second test, the initial position of Treebot was shifted a little bit to the left, as shown in Fig. 7.13(c). This time, branch B was selected by the exploring motion (Fig. 7.13(d)). These results indicate that Treebot tends to choose the closest branch, and thus the selection of a branch is determined by the position of Treebot with the used of the proposed exploring strategy.

7.5 Summary

This chapter presented the development of an autonomous tree-climbing algorithm that enables Treebot to explore and climb autonomously on an unknown and irregularly shaped tree. Inspired by arboreal animals, an algorithm has been proposed to approximate a shape of trees by using limited tactile sensors only which avoid the use of complex sensing equipment such as cameras. The algorithm includes a tree shape approximation method and a motion planning strategy. Numerous experiments have been conducted and the results reveal that the proposed tree shape approximation algorithm along with the exploring strategy can approximate the shape of a tree accurately. Generating an approximated shape of a tree allows Treebot to identify its environment and determine the optimal climbing motion. An associated motion planning algorithm has been proposed, and experimental results show that it

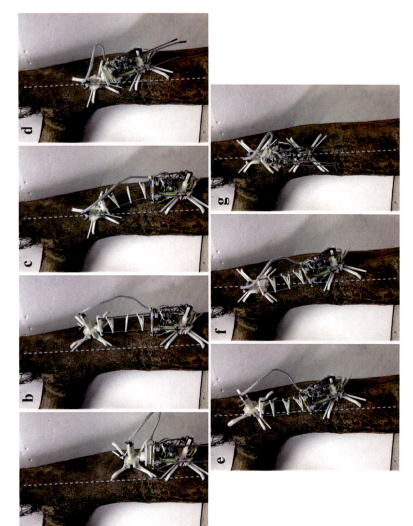

Fig. 7.12 Experimental result of going to optimal path by Strategy 3. (a) Initial position; (b), (e) Exploring motion; (c), (f) Front gripper gripping; (d), (g) Rear gripper gripping.

7.5 Summary

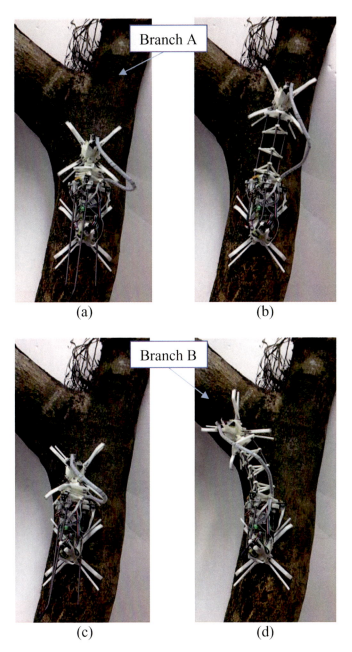

Fig. 7.13 Experiment for climbing a tree with branches. (a) Initial position in the first test; (b) Exploring posture in the first test; (c) Initial position in the second test; (d) Exploring posture in the second test.

successfully guides Treebot to follow the optimal climbing path. It also reveals how tactile sensors can best be used to aid autonomous tree climbing.

In the proposed autonomous climbing algorithm, the selection of a branch is determined passively by the position of Treebot. If Treebot does not move to the desired branch, a manual control is needed to guide it there. To further simplify the control of Treebot, a development of a branch selection function that guides Treebot to climb on a desired branch autonomously will be valuable to develop. It involves the works of global path and motion planning which are going to be discussed in Chapter 8.

Chapter 8
Global Path and Motion Planning[1]

The global motion planning problem for tree climbing is challenging, as trees have an irregular and complex shape. To the best of the author's knowledge, there is no related study that focuses on the global motion planning problem for tree-climbing robot. Aracil [17] proposed a motion planning method to allow the Parallel Climbing Robot to climb a trunk. However, this work merely discussed the local motion planning problem according to local information. There are many motion planning approaches for climbing in artificial structures, such as walls and glass windows [10, 11, 21]. However, these structures are different from trees, and the approaches are thus not suitable for tree-climbing problems.

In the conventional motion planning approach [44], the configuration space (a set of possible transformations that could be applied to the robot [45]) of the problem must be constructed to help solve the problem. However, the formulation of the configuration space is complex as it involves complicated interactions between the environment and the kinematics of the robot. A robot with high degrees of freedom (DOF) and continuous motion requires a high dimensional and huge configuration space, which makes the problem difficult to solve.

This chapter proposes an efficient global motion planning algorithm for tree climbing. It is accomplished by dividing the problem into a path planning and a motion planning problem, which are solved separately to reduce the dimensions of the problem space. In the path planning sub-problem, it is assumed that Treebot is of point size and holonomic, such that its kinematics can be ignored. The aim is to find an optimal path to reach the target position on a 2D manifold. The path planning algorithm includes several constrains to make the path easy for Treebot to follow. As it only considers the 2D manifold of the tree surface, the state space has relatively few dimensions. In addition, an intuitive method is developed to represent the climbing space. It highly simplifies the path planning problem in terms of linear-time complexity. A dynamic programming (DP) algorithm is adopted to find

---

[1] Portions reprinted, with permission from Tin Lun Lam, Guoqing Xu, Huihuan Qian and Yangsheng Xu, "Linear-time Path and Motion Planning Algorithm for a Tree Climbing Robot - TreeBot", Proceedings of the IEEE/RSJ International Conference on Intelligent Robots and Systems. ©[2010] IEEE.

the optimal path according to the specified constrains and requirements. The motion planning sub-problem aims to find an appropriate motion for Treebot that allows it to follow the planned path. An effective strategy for motion planning is proposed, the solution to which can be obtained without any searching effort or additional state space formulation.

The remainder of this chapter is organized as follows. The method of state space formulation is presented in Section 8.1. In Section 8.2, the path planning algorithm is proposed. The motion planning algorithm is presented in Section 8.3. Experimental results are presented in Section 8.4. Finally, summary is given in Section 8.6.

8.1 State Space Formulation

Before working on the path planning sub-problem, the state space to the problem must be formulated. A tree is composed of a trunk and branches. In the proposed algorithm, a trunk is also treated as a branch. It is assumed that the relationship among the branches can be represented by a tree data structure as illustrated in Fig. 8.1. To climb to a target position, a unique sequence of branches must be passed. For example, if the target position is at Branch 8 and the initial position is at branch 1, then there is only one way to go: Branch 1 → Branch 4 → Branch 8. This sequence can be obtained easily by using the backward search method in the tree data structure. This means that in path planning, the climbing space of other non-climbed branches can be neglected, although these branches do need to be considered as obstacles.

The tree surface is discretized by various numbers of points to represent the climbing surface of each branch. The shape of the tree is first decomposed into a number of rings, as shown in Fig. 8.2. The normal direction of a ring is equal to the growth direction of a shape of branch. The distance between each ring takes a certain value such that the rings do not intersect. The shape of each ring is defined by the outer shape of the specified position of the branch, and thus is not necessarily a perfect circle. Finally, each ring is equally discretized by a certain number of points. Each point contains the information about the 3D Cartesian coordinates and the normal vector of the surface of that point. The state space to the problem is arranged in a matrix form with m rows and n columns when a target position is given, as shown in Fig. 8.3. It is composed of the state spaces of the branches to be gone through in sequence.

There are two situations in which Treebot cannot reach a point. The first is when the upper space of a position is not sufficiently large for the robot to pass through, which may occur when the upper space is occupied by other branches. The second is when the gripping surface of a point is concave such that the gripper cannot grip the surface tightly. The state space contains information on such unreachable points.

Information on the shape of the tree can be obtained by several means, such as laser- or vision-based sensing [47, 48]. As the works presented in this chapter are focused on the planning problem, it is assumed that the shape of tree is given. The detail of the sensing and state space conversion problems is not going to be discussed.

8.1 State Space Formulation

Fig. 8.1 Representation of the relationship among branches by using a tree data structure. (a) Real tree structure; (b) Branch relationship as represented by the tree data structure.

Fig. 8.2 Tree surface discretization method.

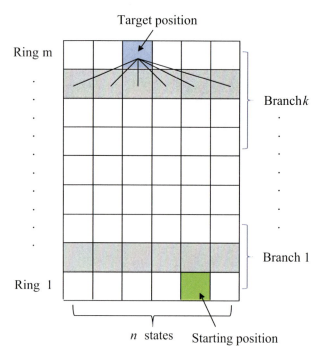

Fig. 8.3 State space representation to the path planning problem.

8.2 Path Planning

Going to a target position and avoiding obstacles are the basic requirements of path planning. In addition, to make a planned path that is easy for Treebot to follow, the planned path should fulfill certain additional requirements. To eliminate the pull out force generated by gravity, Treebot should climb on a upper apex as mentioned in Chapter 7. Furthermore, a shorter path will reduce the robot's energy consumption, and a smoother path will be easier for it to follow. The path should thus be optimized to 1) go directly to the target position, 2) minimize the climbing distance, 3) follow the upper apex of the climbing surface, and 4) avoid obstacles.

8.2.1 Dynamic Programming

Dynamic programming (DP) is an efficient algorithm with a proven ability to find globally optimal solutions to a problem [48]. It works well for discrete states that are difficult to be searched exhaustively. As a result, the DP algorithm is adopted for the path planning problem. The first step in applying DP is to represent the problem in a DP formulation, that is, to identify the **state**, **action**, **action value**, and the **state value** of the problem.

8.2 Path Planning

State $S_{i,j}$: The states of the problem are the discrete points defined in Section 8.1. A state is denoted as $S_{i,j}$ where i and j denotes the row and column of the workspace respectively. The first row represents the starting ring (Ring 1) and the last row represents the ring that contains the target position (Ring m). The elements in each row represent the points in that ring.

Action $S_{i,j} \to S_{i+1,k}$: It is assumed that the target position will not be located on the starting ring, and thus no repeat movement will occur on a ring. Movement can occur to the points on the next ring only. This assumption is reasonable, as climbing motions rarely require moving laterally without moving up or down. This assumption significantly reduces the search space of the problem.

Action value $Q(S_{i,j}, S_{i+1,k})$: The action value is defined as the sum of the reward values:

$$Q(S_{i,j}, S_{i+1,k}) = -D(S_{i,j}, S_{i+1,k}) + a_0 G_{i+1,k} + O_{i+1,k} \quad (8.1)$$

where $D(S_{i,j}, S_{i+1,k})$ represents the Euclidean distance between $S_{i,j}$ and $S_{i+1,k}$. $O_{i,j}$ is the obstacle value. The value is taken as zero if there is no obstacle and $-\infty$ if an obstacle is present. An obstacle means an unreachable point, as defined in Section 8.1. a_0 is a positive scalar value to adjust the weight of $G_{i,j}$ in (8.1). $G_{i,j}$ relates to the amount of the pull-out force generated by the gravity at that point. That the pull-out force is directly proportional to the z component of the normalized surface normal vector $z_{i,j}$ is shown in Fig. 8.4. Hence, the value of $G_{i,j}$ is defined as:

$$G_{i,j} = z_{i,j} - 1 \quad (8.2)$$

where $n_{i,j} \in [-2, 0]$.

State value $V_{i,j}$: Given a target position and the reward values, the state value of each state can be defined. The state value of row m-1 is the distance to the target position, that is,

$$V_{m-1,j} = Q(S_{m-1,j}, S_{m,t}) \quad (8.3)$$

where $S_{m,t}$ denotes the target state.

The state value of the other states can be found by:

$$V_{i,j} = \max\left(V_{i+1,k} + Q(S_{i,j}, S_{i+1,k})\right) \quad (8.4)$$

where $k \in [1, n]$ and $i \in [1, m-1]$.

The next possible states for each state are the points in the next row. As a result, using DP, the computational complexity is only $O(mn^2)$. In fact, the value n is a problem independent value that does not change with the height of the tree. Thus, the computational complexity to solve the problem is only $O(m)$ that can be solved in linear time.

Optimal Path: Once the state value of each state has been defined, the optimal path can be obtained by starting at an arbitrary position or the first row of the state with the maximal state value, and then selecting the state in the next row for which the sum of the state and action values $V_{i+1,k} + Q(S_{i,j}, S_{i+1,k})$ is largest. In that, $S_{i,j}$ and $S_{i+1,k}$ are the current and next state respectively.

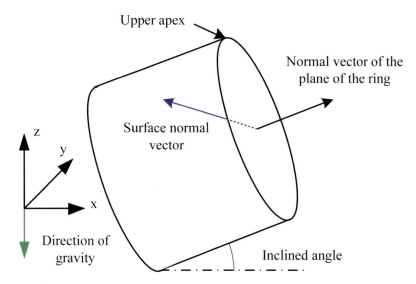

Fig. 8.4 Coordinates and notations for the shape of the tree and the gravity vector.

8.2.2 Dynamic Environment

The structure of a tree will rarely change in a short term. The main reason of updating the information of the environment is that more accurate information on the shape of tree is obtained when Treebot gets closer to a given region. As the calculation of state values in DP is a top-down process, for an ascending motion, a change in environment in the lower part does not affect the state values of the upper part, and only the state values in the lower part need to be modified. The path can then be updated according to the new state values.

Another merit of using DP in this application is that once all the state values of the state space are obtained, the optimal path starting at arbitrary position can easily be obtained in linear time. As Treebot may not go to the target position and orientation exactly due the system error and disturbance, the frequently updated optimal path according to the current position of Treebot is necessary to moderate the path following error.

8.3 Motion Planning

The path planning algorithm generates a 3D path on the manifold of tree surface with high likelihood of success. The next task comes to the motion planning to make Treebot follows the planned path. The ideal solution is that all the steps (front and rear gripper) and the body of robot can place on the planned path. However, finding a motion that keeps both the front and rear grippers and the continuum body

8.3 Motion Planning

on the planned path may not feasible due to the nonholonomic constraints of Treebot's kinematics. It is assumed that the path-following problem has a certain tolerance. This assumption is applicable when the path is planned to avoid obstacles at a certain distance. Searching methods can be applied to find the globally optimal motion sequence to fit the planned path, but this is time consuming. As a result, a computationally efficient strategy to find a near-optimal solution is used rather than exhaustive searching.

8.3.1 Strategy of Motion Planning

It may not be possible to have both grippers and the continuum body on the planned path. As an alternative, either one of the grippers can be placed on the path and then determine the position of the other gripper to minimize the path-following error. A front-gripper-based method, in which all of the steps of the front gripper attach on the planned path, is adopted as it is more intuitive. With this method, the extension motion is used to move the front gripper to the planned path and the contraction motion is used to make the rear gripper adjust the orientation of Treebot to make the next extension motion best fit the planned path. The procedure for the motion-planning scheme is detailed as follows.

8.3.1.1 Path Segmentation

As it is intended that the front gripper is always placed on the planned path, the first task is to determine the target positions of the front gripper on the path. The path between the target positions of the front gripper are defined as path segments of the planned path. As Treebot has a variable length of step, the problem becomes one of determining the length of the continuum body in each climbing gait. As the gripping motion takes time, to climb efficiently, the body should contracts and extends as much as possible so as to minimize the number of gripping motions required. As a result, the length of the contraction motion is set as the minimum admissible length S_{min} while the distance between the target positions of the front gripper is first set as the maximum length of extension of the robot body S_{max}. Once the planned path has been segmented, the next task is to approximate the segment in an arc shape.

8.3.1.2 Arc Fitting

To find an optimal direction of the rear gripper in which the future motion fits the planned path, the path segment should be approximated as an arc as the continuum body has an arc shape. In addition, the target position of the rear gripper should be located near the planned path. As a result, the arc fitting process also considers a rear path segment. The rear path segment is defined as a path below the current position of the front gripper with length S_{min}. The 3D arc fitting method is similar to the method proposed in Chapter 7 with certain modification as illustrated in Fig. 8.5. In the figures, the solid line and dash line represent the path segment and the rear path segment respectively. The fitted arc must pass through the current and target position of the front gripper. Hence, the path segments are transformed as illustrated in Fig. 8.5(b) such that the current position of the front gripper is located at the origin and the target position of the front gripper is lied on z-axis. The plane fitting and 2D arc fitting are then conducted by considering both segments.

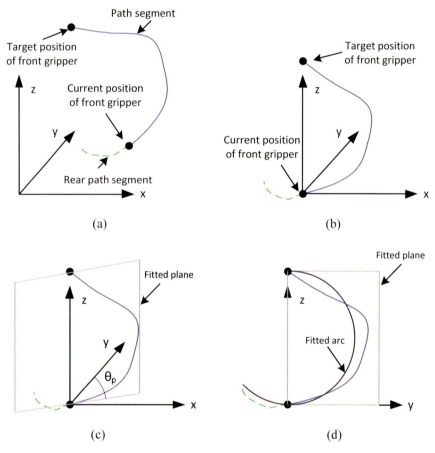

Fig. 8.5 Procedures for 3D arc fitting: (a) A path segment; (b) Transformation; (c) Plane fitting; (d) 2D arc fitting.

8.3.1.3 Optimal Direction of the Rear Gripper

By fitting the arc, the center and radius of the approximated arc can be obtained. The optimal position and direction of the rear gripper can then be determined as illustrated in Fig. 8.6. In the figure, the green dot and arrow represent the optimal position and direction of the rear gripper respectively, and the red arc represents the contraction posture of the continuum body. The rear gripper can be set in this position and direction only if the direction of the front gripper is tangential to the starting point of the approximated arc. However, the direction of the front gripper is uncontrollable when the position of the front gripper is fixed as it is a nonholonomic system. As a result, the optimal position is neglected and the target direction of the rear gripper is set to the optimal direction. With this method, the position of the rear gripper will shift away from the optimal position. This shift will not affect much of the path following result as it is small when compared with the length of the path segment.

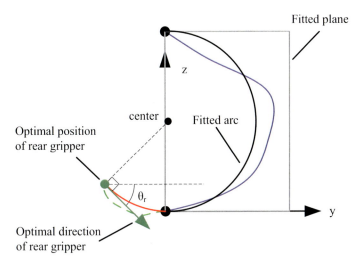

Fig. 8.6 Optimal position and direction of the rear gripper.

The optimal direction of the rear gripper $\mathbf{v}_{rg}$ in the global frame can be obtained by:

$$\mathbf{v}_{rg} = Rot_z(-\theta_z) \, Rot_y(-\theta_y) \, Rot_z(-\theta_p) \begin{bmatrix} 0 \\ \cos\theta_r \\ \sin\theta_r \end{bmatrix} \quad (8.5)$$

where $\theta_r = \tan^{-1}\frac{z_c}{y_c} - \left(\frac{\pi}{2} + \frac{S_{min}}{r}\right)$. $Rot_i(\theta)$ denotes the rotation matrix about i-axis in angle θ where $i \in x, y, z$.

8.3.2 Posture of the Robot

8.3.2.1 Rear Gripper

If the rear gripper is in the optimal direction, it may not capable of setting on the tree surface. It is assumed that the surfaces at the target positions of the rear gripper and the current position of the front gripper have similar properties, as the distance between them is short in the contraction motion. To ensure that the target position of the rear gripper will be on the tree surface, the optimal direction vector is projected on the plane defined by the surface normal to the front gripper position. As a result, to find the appropriate posture of the continuum body for placing the rear gripper, $\mathbf{v}_{rg}$ is first transformed to the front gripper frame so that the center of the front gripper is at the origin, the direction of the front gripper is on the z-axis and the surface normal vector is on the x-axis, as shown in Fig. 8.7. Then, $\mathbf{v}_{rg}$ is projected onto the y-z plane (blue arrow in the figure). Finally, the appropriate posture of the continuum body for placing the rear gripper can be determined as:

$$\phi_r = -\frac{|\theta_{rg}|}{\theta_{rg}}\frac{\pi}{2} \tag{8.6}$$

$$\kappa_r = \frac{|\theta_{rg}|}{S_{min}} \tag{8.7}$$

$$S_r = S_{min} \tag{8.8}$$

where $\tan\theta_{rg} = \frac{z'}{y'}$.

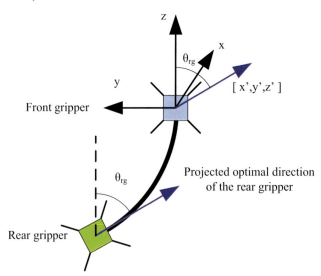

Fig. 8.7 The concept to determine the posture of the continuum body during a contraction motion.

8.3 Motion Planning

8.3.2.2 Front Gripper

The target positions of the front gripper are defined by the path segmentation process. To determine the appropriate posture of the continuum body for placing the front gripper on a target position, the target position is first transformed into the rear gripper frame which is located at the target position of the rear gripper. The target position of the rear gripper can be obtained by (6.7) according to (S_r, ϕ_r, κ_r). In the rear gripper frame, the center of the rear gripper is at the origin and the direction vector of the rear gripper is on the z-axis. The posture of the continuum body to place the front gripper to the target position, (S_f, ϕ_f, κ_f), can then be obtained by (6.6).

8.3.3 Adaptive Path Segmentation

Segmenting a path by a fixed length of path segmentation may induce two problems, a poor quality of the arc fitting result, and the target position for the front gripper is unreachable due to the limitation of the body extension. Those problems can be solved by reducing the length of the path segment adaptively.

8.3.3.1 Quality of the Arc Fitting Result

In some occasions, path segmentation in a constant length may not be fitted closely to an arc as shows in Fig. 8.5(d). The poorly fitted arc degrades the accuracy of the path following as the motion of Treebot is in an arc shape. In this case, reducing the length of the path segment can help improve the fitting quality. The length of the path segment is reduced until the fitness value of the arc is smaller than a positive threshold ε, that is, $max(e_{plane}, e_{arc}) < \varepsilon$.

8.3.3.2 Unreachable Target Position

In the motion planning, even the length of a path segment is defined as the admissible length of extension of the robot body. The front gripper may not capable of going to the target position by two reasons, one is the length of the approximated arc is longer than the admissible length of extension, the second reason is that the target position of the front gripper become farer to the target position after the movement of the rear gripper. The unreachable target position of the front gripper can be detected by checking if $S_f > S_{max}$ before the actual movement. If it is the case, the length of the current path segment reduces and the solution is recalculated until $S_f < S_{max}$.

In summary, Fig. 8.8 shows the procedures of the proposed motion planning strategy. It mainly consists of two loops. The outer loop segments the path in a constant length, and then plans and implements the robot motions. The inner loop fine tunes the length of path segment to get the satisfied solution.

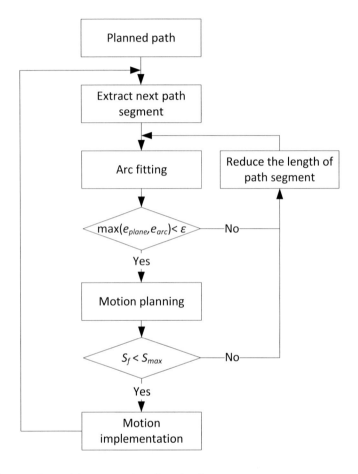

Fig. 8.8 Procedures of the proposed motion planning strategy.

8.4 Simulations

To evaluate the performance of the proposed global path and motion planning algorithm, a tree model that is composed of three branches is constructed. The tree surface is discretized as shown in Fig. 8.9. The rings are marked in different colors to distinguish the branches to which they belong to. The obstacles are marked in magenta.

8.4 Simulations

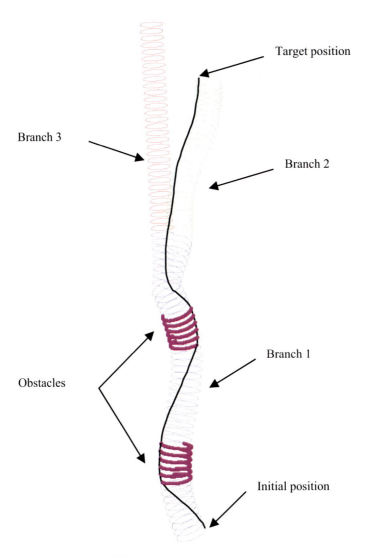

Fig. 8.9 Experimental tree model.

8.4.1 Global Path Planning

To evaluate the global path planning algorithm, a target position is located at the top of Branch 2 and the initial position is located at the bottom of Branch 1, as shown in Fig. 8.9. Fig. 8.10 illustrates the reward value $G_{i,j}$ and $O_{i,j}$ of the selected state space, that is, Branch 1 and Branch 2. In the figure, the hollow regions represent the location of obstacles. The state space arrangement in Fig. 8.10 may not reflect the actual geometric relationship. The planned path generated by the path planning algorithm is shown in Fig. 8.9 and Fig. 8.10 colored in black and grey respectively. In the figures, it can be observed that the planned path successfully reaches the target position by avoiding the obstacles and passing through positions with high reward values.

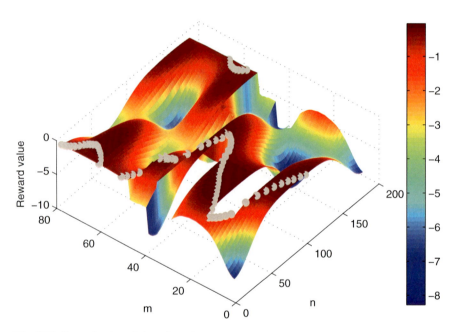

Fig. 8.10 Reward value of the selected state space.

8.4.2 Motion Planning

In the motion planning simulations, it is going to follow the path as shown in Fig. 8.9. Several motion planning simulations by using different planning schemes have been conducted to illustrate the effects of different concerns in the proposed motion planning algorithm. The parameters are set as $S_{max} \approx 100mm$ and $S_{min} \approx 15mm$.

8.4 Simulations

8.4.2.1 Scheme 1

In this motion planning scheme, the planned path is segmented in a constant length. The target direction of the rear gripper is simply equal to the direction of the front gripper. Fig. 8.11(a) and (b) shows the motion planning result in front and left view respectively. To show the motions of Treebot clearly, the obstacles are not displayed in the figures. In the figures, the black line is the planned path, the blue and red arrows indicating the direction and position of the front and rear gripper respectively. The green arc represents the Treebot body in extension motion. The robot requires four climbing gaits to go to the destination. It can be observed that the path following result is not good, several motions (first, second and third gaits) occur far from the planned path. In that, the first and second gait pass through the obstacles, and results in climbing failure. It illustrates that an advanced motion planning algorithm is necessary.

8.4.2.2 Scheme 2

In this scheme, the planned path is also segmented in a constant length, but the target direction of the rear gripper is defined according to the arc fitting method as described above. In that, only the front path segments are considered and the rear path segment is neglected. Fig. 8.12 shows the motion planning result. When compare with Scheme 1, there is a little improvement especially at the second gait as shown in Fig. 8.12(a). However, the result is still unsatisfied. The main reason is that some of the path segments cannot be fitted closely in an arc shape. Table 8.1 shows the length, plane and arc fitness values of each path segment. It can be observed that the fitness values in the first three segments are large which indicates that the path segment cannot be fitted closely as an arc shape. It is thus resulted in a large path following error.

Table 8.1 Length, plane and arc fitness values of the path segments by Scheme 2.

Path segment	Length	Plane fitness value	Arc fitness value
1	101.7056	4.2672	1.6486
2	103.8405	3.0344	3.4747
3	103.7879	3.0126	1.3705
4	102.5579	0.7127	2.6955

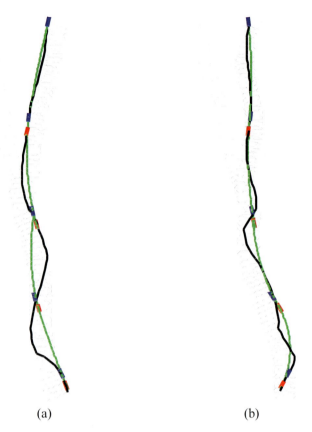

Fig. 8.11 Motion planning results by using Scheme 1. (a) Front view; (b) Left view.

8.4.2.3 Scheme 3

In this scheme, the adaptive path segmentation method is adopted. The plane and arc fitness values are restricted below 1. The arc fitting method still considers the path segments only and the rear path segment is neglected. Fig. 8.13 illustrates the motion planning result and Table 8.2 lists the corresponding fitness values and segment length. The result is much better that that of Scheme 1 and 2. The adaptive length of path segment results in three more gaits required to go to the destination when compare with Scheme 1 and 2. However, minor path following error are still existed in some motions such as the second last and third last gaits.

8.4 Simulations

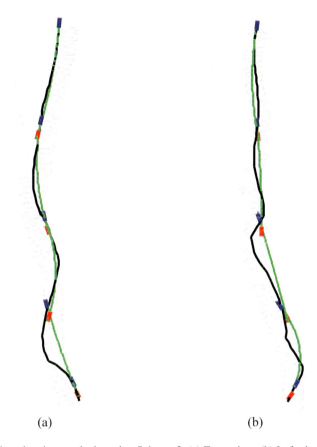

Fig. 8.12 Motion planning results by using Scheme 2. (a) Front view; (b) Left view.

Table 8.2 Length, plane and arc fitness values of the path segments by Scheme 3.

Path segment	Length	Plane fitness value	Arc fitness value
1	41.6051	0.3706	0.5240
2	60.1004	0.6290	0.8152
3	68.0415	0.8390	0.6740
4	43.8508	0.9156	0.8051
5	54.5455	0.7432	0.4643
6	103.8487	0.3411	0.9490
7	39.8998	0.0843	0.1428

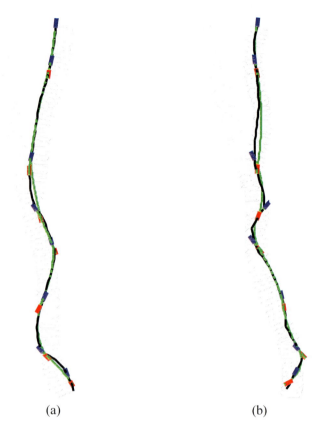

Fig. 8.13 Motion planning results by using Scheme 3. (a) Front view; (b) Left view.

8.4.2.4 Scheme 4

The main different between Scheme 4 and Scheme 3 is that the arc fitting method of Scheme 4 considers both the front and rear path segments as proposed in Section 8.3. Fig. 8.14 illustrates the motion planning result and Table 8.3 lists the corresponding fitness values and segment length. It can be realized that the path following result is further improved when compare with Scheme 3 with two more climbing gaits required. There is no obvious path following error exist.

The comparisons of the motion planning results in different schemes reveal the necessary and significance of each component proposed in Scheme 4, including the target direction of the rear gripper is defined according to the fitted arc, the length of path segment is adaptive to the arc fitting values, and the consideration of both the front and rear path segments in arc fitting.

8.4 Simulations

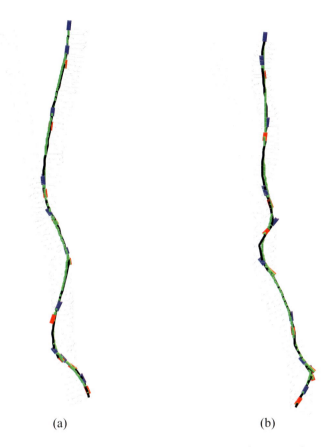

Fig. 8.14 Motion planning results by using Scheme 4. (a) Front view; (b) Left view.

Table 8.3 Length, plane and arc fitness values of the path segments by Scheme 4.

Path segment	Length	Plane fitness value	Arc fitness value
1	32.3859	0.4718	0.5626
2	15.4248	0.9811	0.3195
3	48.1392	0.9993	0.9369
4	67.9364	0.9926	0.5825
5	49.7117	0.9814	0.6281
6	34.0526	0.5615	0.9908
7	71.9228	0.6248	0.8871
8	66.7931	0.6713	0.9098
9	25.5255	0.1236	0.1738

8.5 Experiments

An experiment has been conducted on a tree to evaluate the proposed planning algorithm in a real situation. Fig. 8.15(a) shows the target tree of climbing and the target climbing position. The corresponding tree model which is approximated by a manual measurement along with the path and motion planning results are illustrated in Fig. 8.15(b). In this experiment, Scheme 4 is adopted for the motion planning. The motion planning result shows that Treebot have to take seven climbing gaits to climb to the target position by avoiding the obstacle. Fig. 8.16 shows the actual climbing motions of Treebot according to the planned motions. It can be observed that Treebot climbed along the planned path according to the planned motion and went close to the target position. The position error to the target position was mainly due to the inaccurate modeling of the shape of the tree and the neglect of the deformation of the continuum body due to the gravitational force.

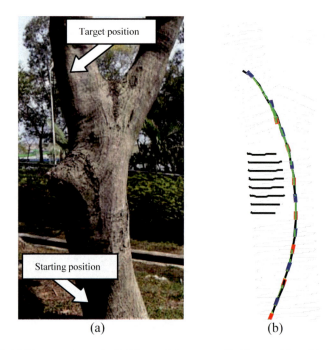

Fig. 8.15 (a) The target tree of climbing and the target climbing position; (b) Approximated tree model and the solutions of path and motion planning.

8.6 Summary

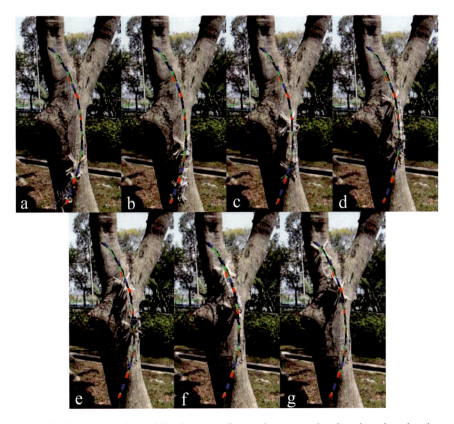

Fig. 8.16 Climbing motions of Treebot according to the proposed path and motion planning algorithm.

8.6 Summary

In summary, this chapter presented a global motion planning strategy for tree climbing that guides Treebot to climb to a target position based on a given global shape of tree. An intuitive method was proposed to represent a climbing space that simplifies the complexity of the problem. A dynamic programming algorithm (DP) is then adopted to find the optimal climbing path that minimizes the climbing effort and avoids obstacles. A path planning solution is thereby obtained in linear-time. A computationally efficient motion planning algorithm was also proposed to guide Treebot to follow the planned path. The performance of the proposed motion planning schemes was compared through simulations. Results revealed that a significant improvement can be made by using the selected motion planning scheme. An experiment of applying the proposed planning algorithm on climbing a real tree was conducted with Treebot to verify the proposed global path and motion planning method in practice. The experimental results revealed that Treebot can climb only close to the target position. The inaccurate result is mainly resulted by the inaccurate modeling of the shape of the tree and the neglect of the deformation of the continuum body due to the gravitational force.

Chapter 9
Conclusions and Future Work

9.1 Conclusion

This book reports a great deal on analysis, mechanical design, motion planning and control of Treebot. To conclude this book, the contributions to the field of tree-climbing robots presented in this book are outlined as follows.

9.1.1 Methodology and Design Principle for Tree-Climbing Robots

A comprehensive study of tree-climbing methodologies in both natural and artificial aspects was undertaken. The major fastening and maneuvering methods for climbing on arboreal environments were introduced and discussed. The fastening methods were categorized as wet adhesion, Van der Waals force, claw penetration, and frictional gripping. As for the maneuvering methods, they were categorized as wave form, body bending, extend-contract, extend-contract with bend, tripod gait, quadruped gait, wheel-driven, and rolling in helical shape. The fastening and maneuvering methods were ranked based on the proposed design principles, i.e., maneuverability, robustness, complexity, adaptiveness, size, and speed. It is found that there is no one method that is the best in all aspects. Each of the methods has it own merit in view of different design principles. The analysis provides a comprehensive reference to help robot designers in selecting the most appropriate climbing methods in designing a tree-climbing robot for specific purpose.

9.1.2 A Novel Tree-Climbing Robot with Distinguish Performance

In this book, the development of a novel tree-climbing robot, Treebot, was presented. It has distinct advantages over conventional tree-climbing robots. It is the world's lightest, smallest and most flexible tree-climbing robot. It successfully overcomes the workspace limitation problem in the conventional tree-climbing robots. Treebot is especially suitable for long time operations, working on the crown of a tree, and performing light duty tasks such as tree inspection, maintenance, pest control and monitor arboreal environment for ecological research. One of the original contributions of this work is the application of the extendable continuum mechanism as a maneuvering system for tree climbing. This opens up a new field of applications for the continuum mechanism. As the inherent compliance of the continuum mechanism gives a certain self adaptiveness to the environment, it is especially suitable for working on unstructured environment such as tree climbing to simplify the control issues. The continuum mechanism used in Treebot adopts a novel driving mechanism which permits superior extensibility. It permits Treebot to have high maneuverability such that the admissible climbing workspace surpasses that of all the state of the art tree-climbing robots. In addition, the proposed extendable continuum mechanism is compact and hence keeps Treebot in lightweight.

Another contribution is the development of the miniature omni-directional tree gripper. The unique mechanical design makes the gripper compact and simple to control. It consumes zero energy in static gripping, which enables Treebot to remain on a tree for a long time. The gripper is also capable of fastening on a wide variety of trees with a wide range of gripping curvatures which have been verified by numerous experiments. This allows Treebot to climb between a large tree trunk and small branches without any change in the gripper settings. It is one of the key elements that Treebot is capable of climbing from a trunk to the crown of a tree as the size of a tree trunk is usually larger than the branches at the crown of the tree in several times. On top of that, the gripper settings have been optimized to generate the maximum gripping force. The experimental results found that the gripper performs well on trees which bark will not peel off easily.

9.1.3 Kinematics and Workspace Analysis

The kinematic model of Treebot was developed and presented in this book, which is actually a kinematic model of a continuum manipulator with straight rods connecting to the ends of the continuum manipulator. It considers the relationship among the length of tendons, posture of the continuum body and the Cartesian coordinates at the end points based on different frames. The workspace analysis has also been conducted to reveal the capability and limitation of the continuum manipulator in terms of locomotion. It offers a complete insight to a robot designer on understanding the admissible climbing motions of a robot that uses continuum mechanism for maneuvering. More importantly, the study of the kinematics and workspace of a continuum mechanism provides crucial information for motion planning and autonomous control in the type of robots that adopts continuum mechanism for maneuvering.

9.1 Conclusion

9.1.4 Autonomous Climbing Strategy in an Unknown Environment

Inspired by arboreal animals, an autonomous tree-climbing algorithm was developed that enables Treebot to explore and climb on an unknown shape of tree autonomously. A computational efficient exploration approach has been proposed which uses limited tactile sensors instead of complex sensing equipment such as cameras. A 3D arc fitting algorithm was proposed to reconstruct the shape of tree according to the data acquired by the tactile sensors. The experimental results reveal that the proposed exploring strategy along with the tree shape approximation algorithm can approximate the shape of trees correctly. The approximated shape of tree permits Treebot to identify the environment and helps determine the optimal climbing position, i.e., the position above the centerline of the tree to minimize the tendency of toppling sideways. An associated motion planning algorithm was thus proposed to guide Treebot going to the optimal position. Several feasible motion planning strategies were discussed. Experimental results show that the integration of the works make Treebot to follow the optimal path successfully. The study and the experiments also reveal how best tactile sensors can be used to aid autonomous tree climbing.

9.1.5 Global Path and Motion Planning on Climbing Irregularly Shaped Trees

A global path and motion planning strategy for tree climbing was developed to guide Treebot to climb to a target position based on a given shape of tree. To the best of the author's knowledge, this is the first study to solve the global motion planning problem for tree climbing. The path and motion planning problem are solved by dividing it into a path planning and a motion planning sub-problem, which are solved separately to reduce the dimensions of the problem space. An intuitive method has been proposed to represent a climbing space in a tree data structure that highly simplifies the complexity to the path planning problem. A dynamic programming (DP) algorithm is adopted to find an optimal climbing path that minimizes the climbing effort and avoids obstacles. A path planning solution is thereby obtained in linear-time. A computationally efficient motion planning algorithm was also proposed to guide Treebot to follow the planned path. It is achieved by using an adaptive segmentation technique on the planned path and an intuitive front-gripper-based motion planning strategy. The performance of the proposed motion planning method was verified by both simulation and experimental results.

9.2 Future Research Directions

Based on the works we have achieved, we look beyond our current study and propose some future research issues. The followings are some possible applications, improvements and extension of the works.

9.2.1 Fastening Mechanism

The current fastening mechanism has four contact points to the substrate, which is mainly designed for lifting miniature robots. To extend the use of the mechanism to large scale robots, or capable of equipping heavier objects, the mechanism has to be improved so as to generate larger fastening force. The improved fastening mechanism also has potential to apply on other fields such as rock climbing, or acted as a shoe for astronauts to help fasten them on the ground of a planet with near-zero gravity. In order to optimize the fastening mechanism, the relationship among the claw insert angle, stiffness of the substrate, direction of acting force, and fastening force generated in different direction should be formulated. Provancher [35] conducted a set of experiments in two-dimensional plane only. In addition, it provides data in certain set of experimental setups only and no equation in modeling the relationship is presented. As a result, to optimize the settings of the fastening mechanism, a comprehensive study and analysis of the interaction between claw and substrate in three-dimensional space should be conducted and a model in representing the relationship should be formulated in the future. On the other hand, increasing the number of contact points can reduce the load shared on each claw. It can then increase the maximal fastening force of the gripper. In order to keep the mechanism in lightweight and simple in control by minimizing the use of actuators, a certain level of compliance on the claws should be adopted in the mechanism. As a result, a proper mechanism to increase the number of contact points and at the same time keep the optimal insert angle by utilizing compliance is worth to be investigated in the future.

9.2.2 Continuum Mechanism

One of the distinct features of the continuum mechanism is that it has a certain level of compliance so as to adapt to unstructured environment. However, the compliance is also a drawback in view of position control as the posture of the continuum mechanism will be affected by external force such as gravitational force. The kinematic model of the continuum mechanism neglects the deflection and hence it is not accurate enough to describe the actual posture of the continuum mechanism in practice. The position control of the continuum mechanism under different external load becomes a challenging problem. In order to approximate the posture of the continuum mechanism accurately which is especially useful for motion planning in autonomous control, the dynamics of the continuum mechanism including the deflection due to the exertion of external forces must be modeled. The analysis of the dynamics of the continuum mechanism is also useful for selecting the optimal parameters of the setting of the mechanism such as the stiffness of tendons and the power of motors so as to fit for different application.

On the other hand, we are seeking the possibility of applying the proposed extendible continuum manipulator technologies on other field robots, such as pipe robot, wall climbing robot and pole climbing robot or even applying on other fields of application. We believe that these technologies are suitable in overcoming the engineering challenges for the autonomous field robot for similar system as Treebot.

9.2.3 Map Building and Localization

We have proposed a global path and motion planning algorithm for tree climbing. It is only the first step towards full implementation. To allow autonomous climbing globally, a global shape of a tree must be obtained. An appropriate method for obtaining a map to describe a tree surface should thus be developed in the future. A 3D tree map which consisted of the 3D shape of tree and the surface images is useful for health inspection of trees and conducting researches on the area of agriculture and forestry. After obtaining the 3D tree map, users can locate where they want Treebot to go on a 3D tree map and Treebot can go to the target position on a tree autonomously. It highly simplifies the work of manipulating Treebot. In addition to map building, a method for locating the actual position of the robot on the tree is necessary to compensate the path following error caused by the environment uncertainty and modeling error. There are mainly two technical challenges, map building and robot localization which can be achieved by utilizing simultaneous localization and mapping (SLAM) technique. In view of the similarity of the color and texture of the surface of trees, it is difficult to use passive stereo imaging technique to obtain the shape in 3D. It is suggested using active type RGBD camera such as Kinect to obtain the 3D shape and the color image simultaneously. Since it uses active light to obtain the 3D shape, it also suffers from fewer disturbances due to the outdoor lighting condition.

Appendix A
Derivation of Equations

A.1 Kinematics of the Continuum Body

A.1.1 Inverse Kinematics

Initial position of tendon 1: $(-d, 0)$
Initial position of tendon 2: $\left(\frac{1}{2}d, -\frac{\sqrt{3}}{2}d\right)$
Initial position of tendon 3: $\left(\frac{1}{2}d, \frac{\sqrt{3}}{2}d\right)$

After transformation to the new direction of bend (rotation of about z-axis), the x coordinates of the initial position of tendons become,

$$T_{1x} = -d\cos\phi \tag{A.1}$$

$$T_{2x} = \frac{1}{2}d\cos\phi - \frac{\sqrt{3}}{2}d\sin\phi = d\sin\left(\frac{\pi}{6} - \phi\right) \tag{A.2}$$

$$T_{3x} = \frac{1}{2}d\cos\phi + \frac{\sqrt{3}}{2}d\sin\phi = d\sin\left(\frac{\pi}{6} + \phi\right) \tag{A.3}$$

As each tendon should have same θ,

$$l_i = (r - T_{ix})\theta \tag{A.4}$$

Since $\theta = \kappa S$ and $r = 1/\kappa$, (A.4) can be rewritten as,

$$l_i = \left(\frac{1}{\kappa} - T_{ix}\right)\kappa S = (1 - \kappa T_{ix})S \tag{A.5}$$

As a result, the inverse kinematics becomes,

$$l_1 = S(1 + \kappa d \cos\phi) \tag{A.6}$$

$$l_2 = S\left(1 - \kappa d \sin\left(\frac{\pi}{6} - \phi\right)\right) \tag{A.7}$$

$$l_3 = S\left(1 - \kappa d \sin\left(\frac{\pi}{6} + \phi\right)\right) \tag{A.8}$$

A.1.2 Forward Kinematics

To find the forward kinematics, sub. (A.6) into (A.7):

$$l_2 = \frac{l_1}{(1 + \kappa d \cos\phi)}\left(1 - \kappa d \sin\left(\frac{\pi}{6} - \phi\right)\right)$$

$$\Rightarrow \kappa = \frac{l_1 - l_2}{d\left(l_1 \sin\left(\frac{\pi}{6} - \phi\right) + l_2 \cos\phi\right)} \tag{A.9}$$

Sub. (A.6) into (A.8):

$$l_3 = \frac{l_1}{(1 + \kappa d \cos\phi)}\left(1 - \kappa d \sin\left(\frac{\pi}{6} + \phi\right)\right)$$

$$\Rightarrow \kappa = \frac{l_1 - l_3}{d\left(l_1 \sin\left(\frac{\pi}{6} + \phi\right) + l_3 \cos\phi\right)} \tag{A.10}$$

Combine (A.9) and (A.10):

$$\frac{l_1 - l_2}{d\left(l_1 \sin\left(\frac{\pi}{6} - \phi\right) + l_2 \cos\phi\right)} = \frac{l_1 - l_3}{d\left(l_1 \sin\left(\frac{\pi}{6} + \phi\right) + l_3 \cos\phi\right)}$$

$$(l_1 - l_2)\left(l_1 \sin\left(\frac{\pi}{6} + \phi\right) + l_3 \cos\phi\right)$$
$$= (l_1 - l_3)\left(l_1 \sin\left(\frac{\pi}{6} - \phi\right) + l_2 \cos\phi\right)$$

Since $\sin\left(\frac{\pi}{6} \pm \phi\right) = \sin\left(\frac{\pi}{6}\right)\cos\phi \pm \cos\left(\frac{\pi}{6}\right)\sin\phi = \frac{1}{2}\cos\phi \pm \frac{\sqrt{3}}{2}\sin\phi$,

$$(l_1 - l_2)\left(\left(\frac{l_1}{2} + l_3\right)\cos\phi + l_1\frac{\sqrt{3}}{2}\sin\phi\right)$$

$$= (l_1 - l_3)\left(\left(\frac{l_1}{2} + l_2\right)\cos\phi - l_1\frac{\sqrt{3}}{2}\sin\phi\right)$$

$$(l_1 - l_2)\left((l_1 + 2l_3)\cos\phi + \sqrt{3}l_1 \sin\phi\right)$$

A.1 Kinematics of the Continuum Body

$$= (l_1 - l_3)\left((l_1 + 2l_2)\cos\phi - \sqrt{3}l_1 \sin\phi\right)$$

$$\left[\sqrt{3}l_1(l_1 - l_2) + \sqrt{3}l_1(l_1 - l_3)\right]\sin\phi$$

$$= [(l_1 - l_3)(l_1 + 2l_2) - (l_1 - l_2)(l_1 + 2l_3)]\cos\phi$$

$$\sqrt{3}l_1(2l_1 - l_2 - l_3)\sin\phi = 3l_1(l_2 - l_3)\cos\phi$$

$$\frac{\sin\phi}{\cos\phi} = \frac{3(l_2 - l_3)}{\sqrt{3}(2l_1 - l_2 - l_3)}$$

$$\tan\phi = \frac{\sqrt{3}(l_2 - l_3)}{(2l_1 - l_2 - l_3)} \tag{A.11}$$

$$\Rightarrow \phi = \tan^{-1}\frac{\sqrt{3}(l_2 - l_3)}{(2l_1 - l_2 - l_3)} \tag{A.12}$$

Eq. (A.9) can be rewritten as:

$$\kappa = \frac{l_1 - l_2}{d\left(l_1\left(\frac{1}{2}\cos\phi - \frac{\sqrt{3}}{2}\sin\phi\right) + l_2\cos\phi\right)}$$

$$= \frac{2(l_1 - l_2)}{d\left(l_1(\cos\phi - \sqrt{3}\sin\phi) + 2l_2\cos\phi\right)}$$

$$= \frac{2(l_1 - l_2)}{d\left((l_1 + 2l_2)\cos\phi - \sqrt{3}l_1\sin\phi\right)}$$

$$= \frac{2(l_1 - l_2)}{d\cos\phi\left((l_1 + 2l_2) - \sqrt{3}l_1\tan\phi\right)} \tag{A.13}$$

Since $\cos\phi = \frac{1}{\sqrt{1+\tan^2\phi}}$,

$$\kappa = \frac{2(l_1 - l_2)\sqrt{1 + \tan^2\phi}}{d\left((l_1 + 2l_2) - \sqrt{3}l_1\tan\phi\right)} \tag{A.14}$$

Sub. (A.11) into (A.14):

$$\kappa = \frac{2(l_1 - l_2)\sqrt{1 + \left(\frac{\sqrt{3}(l_2-l_3)}{(2l_1-l_2-l_3)}\right)^2}}{d\left((l_1 + 2l_2) - \sqrt{3}l_1\frac{\sqrt{3}(l_2-l_3)}{(2l_1-l_2-l_3)}\right)}$$

$$= \frac{2(l_1 - l_2)\sqrt{(2l_1 - l_2 - l_3)^2 + (\sqrt{3}(l_2 - l_3))^2}}{d((2l_1 - l_2 - l_3)(l_1 + 2l_2) - 3l_1(l_2 - l_3))}$$

$$= \frac{2\sqrt{l_1^2 + l_2^2 + l_3^2 - l_1 l_2 - l_2 l_3 - l_1 l_3}}{d(l_1 + l_2 + l_3)} \tag{A.15}$$

Sub. (A.14) into (A.6):

$$S = \frac{l_1}{\left(1 + d\dfrac{2(l_1-l_2)\sqrt{1+\tan^2\phi}}{d\left((l_1+2l_2)-\sqrt{3}l_1\tan\phi\right)}\dfrac{1}{\sqrt{1+\tan^2\phi}}\right)}$$

$$= \frac{l_1}{\left(1 + \dfrac{2(l_1-l_2)}{\left((l_1+2l_2)-\sqrt{3}l_1\tan\phi\right)}\right)} \qquad (A.16)$$

Sub. (A.11) into (A.16):

$$S = \frac{l_1}{\left(1 + \dfrac{2(l_1-l_2)}{\left((l_1+2l_2)-\sqrt{3}l_1\dfrac{\sqrt{3}(l_2-l_3)}{(2l_1-l_2-l_3)}\right)}\right)}$$

$$= \frac{l_1}{\left(1 + \dfrac{2(l_1-l_2)(2l_1-l_2-l_3)}{((2l_1-l_2-l_3)(l_1+2l_2)-3l_1(l_2-l_3))}\right)}$$

$$= \frac{l_1}{\left(1 + (2l_1-l_2-l_3)\dfrac{2(l_1-l_2)}{((2l_1-l_2-l_3)(l_1+2l_2)-3l_1(l_2-l_3))}\right)}$$

$$= \frac{l_1}{\left(1 + \dfrac{(2l_1-l_2-l_3)}{(l_1+l_2+l_3)}\right)}$$

$$= \frac{l_1(l_1+l_2+l_3)}{((l_1+l_2+l_3)+(2l_1-l_2-l_3))}$$

$$= \frac{(l_1+l_2+l_3)}{3} \qquad (A.17)$$

Jones [30] introduces a kinematic model for a general class of continuum robot by different approaches. But it is found that this model is same as the model we have divided. It can bee seen that the equations are similar except the difference of ϕ is $\pi/2$.

A.1.3 Mapping between the Posture and the Cartesian Coordinate

$(S, \kappa, \phi) \leftarrow f(x_t, y_t, z_t)$:
According to Fig. A.1(a), ϕ can be determined by:

$$\phi = \tan^{-1}\frac{y_t}{x_t}$$

To find S and κ, it is first rotate the virtual tendon to x-z plane (refer to Fig. A.1(b)). Then,

$$\theta_1 = \tan^{-1}\frac{z_t}{x_t'}$$

where $x_t' = x_t\cos\phi + y_t\sin\phi$.

A.1 Kinematics of the Continuum Body

Once θ_1 is obtained, the radius of bend r can be found by:

$$r = \frac{\sqrt{x_t'^2 + z_t'^2}}{2\cos\theta_1}$$

Hence,

$$\kappa = \frac{2\cos\theta_1}{\sqrt{x_t'^2 + z_t'^2}} = \frac{2\cos\left(\tan^{-1}\frac{z_t}{x_t'}\right)}{\sqrt{x_t'^2 + z_t'^2}} = \frac{2\sqrt{\frac{1}{1+\left(\frac{z_t}{x_t'}\right)^2}}}{\sqrt{x_t'^2 + z_t'^2}} = \frac{2x_t'}{x_t'^2 + z_t'^2}$$

as $\cos(\tan^{-1} x) = \frac{1}{\sqrt{1+x^2}}$. Refer to Fig. A.1(b)),

$$\theta = 2\left(\frac{\pi}{2} - \theta_1\right) = 2\left(\frac{\pi}{2} - \tan^{-1}\frac{z_t}{x_t'}\right)$$

$$\tan\left(\frac{\pi}{2} - \frac{\theta}{2}\right) = \frac{z_t}{x_t'}$$

$$\cot\frac{\theta}{2} = \frac{z_t}{x_t'}$$

$$\tan\frac{\theta}{2} = \frac{x_t'}{z_t}$$

$$\theta = 2\tan^{-1}\frac{x_t'}{z_t}$$

Since $S = \theta/\kappa$,

$$S = \frac{x_t'^2 + z_t'^2}{x_t'} \tan^{-1}\frac{x_t'}{z_t}$$

$(x_t, y_t, z_t) \leftarrow f(S, \kappa, \phi)$:
According to Fig. A.1,

$$\begin{bmatrix} x'_t \\ 0 \\ z_t \end{bmatrix} = \begin{bmatrix} r - r\cos\theta \\ 0 \\ r\sin\theta \end{bmatrix} = r\begin{bmatrix} 1 - \cos\theta \\ 0 \\ \sin\theta \end{bmatrix}$$

Then,

$$\begin{bmatrix} x_t \\ y_t \\ z_t \end{bmatrix} = Rot_z(\phi)\begin{bmatrix} x'_t \\ 0 \\ z_t \end{bmatrix} = Rot_z(\phi)\, r \begin{bmatrix} 1 - \cos\theta \\ 0 \\ \sin\theta \end{bmatrix}$$

$$= \frac{1}{\kappa}\begin{bmatrix} [1 - \cos(\kappa S)]\cos\phi \\ [1 - \cos(\kappa S)]\sin\phi \\ \sin(\kappa S) \end{bmatrix}$$

A Derivation of Equations

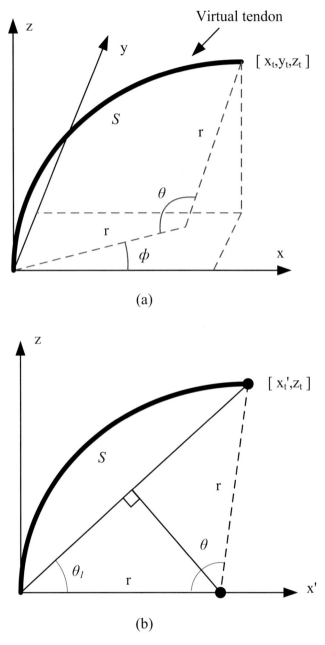

Fig. A.1 Notations of the continuum manipulator.

A.2 Kinematics of Treebot

A.2.1 Mapping between the Posture and the Cartesian Coordinate

A.2.1.1 Rear Gripper Frame

$\left({}^r x_f, {}^r y_f, {}^r z_f \right) \leftarrow f(S, \kappa, \phi)$:

$$\begin{bmatrix} {}^r x'_f \\ 0 \\ {}^r z_f \end{bmatrix} = r \begin{bmatrix} 1 - \cos\theta \\ 0 \\ \sin\theta \end{bmatrix} + l_r \begin{bmatrix} 0 \\ 0 \\ 1 \end{bmatrix} + l_f \begin{bmatrix} \sin\theta \\ 0 \\ \cos\theta \end{bmatrix}$$

$$= \begin{bmatrix} r(1 - \cos\theta) + l_f \sin\theta \\ 0 \\ r \sin\theta + l_r + l_f \cos\theta \end{bmatrix} \quad (A.18)$$

$$\begin{bmatrix} {}^r x_f \\ {}^r y_f \\ {}^r z_f \end{bmatrix} = Rot_z(\phi) \begin{bmatrix} {}^r x'_f \\ 0 \\ {}^r z_f \end{bmatrix} = Rot_z(\phi) \begin{bmatrix} r(1 - \cos\theta) + l_f \sin\theta \\ 0 \\ r \sin\theta + l_r + l_f \cos\theta \end{bmatrix}$$

$$= \begin{bmatrix} \left(\frac{1}{\kappa}[1 - \cos(\kappa S)] + l_f \sin(\kappa S) \right) \cos\phi \\ \left(\frac{1}{\kappa}[1 - \cos(\kappa S)] + l_f \sin(\kappa S) \right) \sin\phi \\ \frac{1}{\kappa} \sin(\kappa S) + l_f \cos(\kappa S) + l_r \end{bmatrix}$$

$(S, \kappa, \phi) \leftarrow f\left({}^r x_f, {}^r y_f, {}^r z_f \right)$:

$$\phi = \tan^{-1} \frac{{}^r y_f}{{}^r x_f}$$

To find S and κ, it is first rotate the virtual tendon to x-z plane. In addition, transform l_r in z-axis (refer to Fig. A.2), thus, ${}^r z'_f = {}^r z_f - l_r$ and ${}^r x'_f = {}^r x_f \cos\phi + {}^r y_f \sin\phi$. Then,

$${}^r x'_f = r(1 - \cos\theta) + l_f \sin\theta \quad (A.19)$$
$${}^r z'_f = r(\sin\theta) + l_f \cos\theta \quad (A.20)$$

Reform (A.20):

$$r = \frac{{}^r z'_f - l_f \cos\theta}{\sin\theta} \quad (A.21)$$

Sub. (A.20) into (A.19):

$$^r x'_f = \frac{^r z'_f - l_f \cos\theta}{\sin\theta}(1-\cos\theta) + l_f \sin\theta$$

$$^r x'_f \sin\theta = \left(^r z'_f - l_f \cos\theta\right)(1-\cos\theta) + l_f \sin^2\theta$$
$$= {}^r z'_f - \left(^r z'_f + l_f\right)\cos\theta + l_f\cos^2\theta + l_f\sin^2\theta$$

$$\frac{\sin\theta}{1-\cos\theta} = \frac{{}^r z'_f + l_f}{{}^r x'_f} = u \qquad (A.22)$$

Let $\tan\theta = t$, $\sin\theta = \frac{t}{\sqrt{1+t^2}}$ and $\cos\theta = \frac{1}{\sqrt{1+t^2}}$, then,

$$\frac{\sin\theta}{1-\cos\theta} = \frac{\frac{t}{\sqrt{1+t^2}}}{1-\frac{1}{\sqrt{1+t^2}}} = \frac{t}{\sqrt{1+t^2}-1} = \frac{\sqrt{1+t^2}+1}{t} = u$$

$$\Rightarrow (ut-1)^2 = 1+t^2$$
$$u^2 t^2 - 2ut + 1 = 1 + t^2$$
$$t\left[(u^2-1)t - 2u\right] = 0$$

$$t = \frac{2u}{u^2-1} \qquad (A.23)$$

Sub. (A.22) into (A.23):

$$t = \frac{2\left(\frac{{}^r z'_f + l_f}{{}^r x'_f}\right)}{\left(\frac{{}^r z'_f + l_f}{{}^r x'_f}\right)^2 - 1} = \frac{2\, {}^r x'_f \left(^r z'_f + l_f\right)}{\left(^r z'_f + l_f\right)^2 - {}^r x'^2_f}$$

$$\Rightarrow \theta = \tan^{-1}\frac{2\, {}^r x'_f \left(^r z'_f + l_f\right)}{\left(^r z'_f + l_f\right)^2 - {}^r x'^2_f}$$

From (A.21):

$$r = \frac{{}^r z'_f - l_f \cos\theta}{\sin\theta} = \frac{{}^r z'_f - l_f \frac{1}{\sqrt{1+t^2}}}{\frac{t}{\sqrt{1+t^2}}} = \frac{{}^r z'_f \sqrt{1+t^2} - l_f}{t} \qquad (A.24)$$

A.2 Kinematics of Treebot

Sub. (A.23) into (A.24):

$$r = \frac{{}^r z'_f \sqrt{1+\left(\frac{2u}{u^2-1}\right)^2}-l_f}{\frac{2u}{u^2-1}} = \frac{{}^r z'_f \frac{u^2+1}{u^2-1}-l_f}{\frac{2u}{u^2-1}}$$

$$= \frac{(u^2+1){}^r z'_f - (u^2-1)l_f}{2u}$$

$$= \frac{\left({}^r z'_f - l_f\right)u^2 + {}^r z'_f + l_f}{2u} \tag{A.25}$$

Sub. (A.22) into (A.25):

$$r = \frac{\left({}^r z'_f - l_f\right)\left(\frac{{}^r z'_f + l_f}{{}^r x'_f}\right)^2 + \left({}^r z'_f + l_f\right)}{2\frac{{}^r z'_f + l_f}{{}^r x'_f}}$$

$$= \frac{\left({}^r z'_f - l_f\right)\left({}^r z'_f + l_f\right)^2 + {}^r x'^2_f\left({}^r z'_f + l_f\right)}{2x'_t\left({}^r z'_f + l_f\right)}$$

$$= \frac{{}^r z'^2_f - l_f^2 + {}^r x'^2_f}{2\,{}^r x'_f}$$

Hence,

$$\kappa = \frac{1}{r} = \frac{2\,{}^r x'_f}{{}^r z'^2_f - l_f^2 + {}^r x'^2_f}$$

Finally,

$$S = r\theta = \frac{{}^r z'^2_f - l_f^2 + {}^r x'^2_f}{2\,{}^r x'_f}\tan^{-1}\frac{2\,{}^r x'_f\left({}^r z'_f + l_f\right)}{\left({}^r z'_f + l_f\right)^2 - {}^r x'^2_f}$$

A.2.1.2 Front Gripper Frame

For the case of the front gripper based, the derivation is similar to the rear gripper based but mirror about z-axis. Hence the difference are: ${}^r z_f \to -{}^f z_r$, $l_f \to l_r$ and $l_r \to l_f$.

As a result, $({}^f x_r, {}^f y_r, {}^f z_r) \leftarrow f(S, \kappa, \phi)$:

$$\begin{bmatrix} {}^f x_r \\ {}^f y_r \\ {}^f z_r \end{bmatrix} = \begin{bmatrix} \left(\frac{1}{\kappa}[1-\cos(\kappa S)]+l_r\sin(\kappa S)\right)\cos\phi \\ \left(\frac{1}{\kappa}[1-\cos(\kappa S)]+l_r\sin(\kappa S)\right)\sin\phi \\ -\left(\frac{1}{\kappa}\sin(\kappa S)+l_r\cos(\kappa S)+l_f\right) \end{bmatrix}$$

A Derivation of Equations

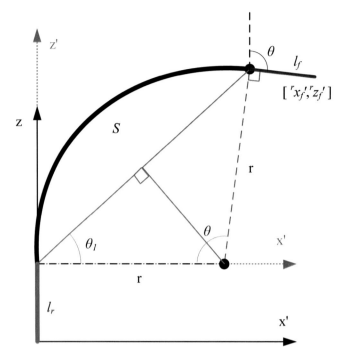

Fig. A.2 Notations of Treebot.

$(S, \kappa, \phi) \leftarrow f\left({}^f x_r, {}^f y_r, {}^f z_r\right)$:

$$\begin{bmatrix} S \\ \kappa \\ \phi \end{bmatrix} = \begin{bmatrix} \frac{1}{\kappa}\tan^{-1}\left(\frac{2^f\hat{x}_r\left({}^f\hat{z}_r+l_r\right)}{\left({}^f\hat{z}_r+l_r\right)^2 - {}^f\hat{x}_r^2}\right) \\ \frac{2^f\hat{x}_r}{{}^f\hat{x}_r^2 + {}^f\hat{z}_r^2 - l_r^2} \\ \tan^{-1}\frac{{}^f y_r}{{}^f x_r} \end{bmatrix}$$

where ${}^f\hat{x}_r = {}^f x_r \cos\phi + {}^f y_r \sin\phi$ and ${}^f\hat{z}_r = -{}^f z_r - l_f$.

References

1. Kim, S., Spenko, M., Trujillo, S., Heyneman, B., Santos, D., Cutkosky, M.R.: Smooth Vertical Surface Climbing With Directional Adhesion. IEEE Transactions on Robotics 24(1), 65–74 (2008)
2. Prahlad, H., Pelrine, R., Stanford, S., Marlow, J., Kornbluh, R.: Electroadhesive Robots-Wall Climbing Robots Enabled by a Novel, Robust, and Electrically Controllable Adhesion Technology. In: Pasadena, C.A. (ed.) Proceedings of the IEEE International Conference on Robotics and Automation, Pasadena, CA, USA, May 19-23, pp. 3028–3033 (2008)
3. Aksak, B., Murphy, M.P., Sitti, M.: Gecko inspired micro-fibrillar adhesives for wall climbing robots on micro/nanoscale rough surfaces. In: Proceedings of the IEEE International Conference on Robotics and Automation, Pasadena, CA, USA, May 19-23, pp. 3058–3063 (2008)
4. Xu, D., Gao, X., Wu, X., Fan, N., Li, K., Kikuchi, K.: Suction Ability Analyses of a Novel Wall Climbing Robot. In: Proceedings of the IEEE International Conference on Robotics and Biomimetics, Kunming, China, December 17-20, pp. 1506–1511 (2006)
5. Shen, W., Gu, J., Shen, Y.: Permanent Magnetic System Design for the Wall-climbing Robot. In: Proceedings of the IEEE International Conference on Mechatronics and Automation, Niagara Falls, Canada (July 2005)
6. Segal, S.J., Virost, S., Provancher, W.R.: ROCR: Dynamic Vertical Wall Climbing with a Pendular Two-Link Mass-Shifting Robot. In: Proceedings of the IEEE International Conference on Robotics and Automation, Pasadena, CA, USA, May 19-23, pp. 3040–3045 (2008)
7. Murphy, M., Sitti, M.: Waalbot: An Agile Small-Scale Wall Climbing Robot Utilizing Dry Elastomer Adhesives. IEEE/ASME Transactions on Mechatronics 12(3) (June 2007)
8. Zhang, H., Zhang, J., Zong, G.: Effective pneumatic scheme and control strategy of a climbing robot for class wall cleaning on high-rise buildings. International Journal of Advanced Robotic Systems 3(2), 183–190 (2006)
9. Sintov, A., Avramovich, T., Shapiro, A.: Design and motion planning of an autonomous climbing robot with claws. Robotics and Autonomous Systems 59(11), 1008–1019 (2011)
10. Balaguer, C., Gimenez, A., Pastor, J.M., Padron, V.M., Abderrahim, M.: A climbing autonomous robot for inspection applications in 3D complex environments. Robotica 18(3), 287–297 (2000)

11. Yoon, Y., Rus, D.: Shady3D: A Robot that Climbs 3D Trusses. In: Proceedings of the IEEE International Conference on Robotics and Automation, Roma, Italy, April 10-14, pp. 4071–4076 (2007)
12. Mahmoud, T., Ali, M., Lino, M., de Anibal, A.T.: 3DCLIMBER: A climbing robot for inspection of 3D human made structures. In: Proceedings of the IEEE International Conference on Intelligent Robots and Systems, Nice, France, September 22-26, pp. 4130–4135 (2008)
13. Baghani, A., Ahmadabadi, M., Harati, A.: Kinematics Modelling of a Wheel-Based Pole Climbing Robot (UT-PCR). In: Proceedings of the IEEE International Conference on Robotics and Automation, Barcelona, Spain, April 18-22, pp. 2099–2104 (2005)
14. Sattar, T.P., Rodriguez, H.L., Bridge, B.: Climbing ring robot for inspection of offshore wind turbines. International Journal of Industrial Robot 36(4), 326–330 (2009)
15. Kushihashi, Y., et al.: Development of Tree Climbing and Pruning Robot, Woody-1-Simplification of Control using adjust Function of Grasping Power. In: Proceedings of JSME Conference on Robotics and Mechatronics, pp. 1A1–E08 (2006) (in Japanese)
16. Kawasaki, H., et al.: Novel climbing method of pruning robot. In: Proceedings of the SICE Annual Conference, Tokyo, pp. 160–163 (2008)
17. Aracil, R., Saltarn, R.J., Reinoso, O.: A climbing parallel robot: a robot to climb along tubular and metallic structures. IEEE Robotics and Automation Magazine 13(1), 16–22 (2006)
18. Spenko, M.J., Haynes, G.C., Saunders, J.A., Cutkosky, M.R., Rizzi, A.A.: Biologically Inspired Climbing with a Hexapedal Robot. Journal of Field Robotics 25(4-5), 223–242 (2008)
19. Haynes, G., et al.: Rapid Pole Climbing with a Quadrupedal Robot. In: Proceedings of the IEEE International Conference on Robotics and Automation, Kobe, Japan, May 12-17, pp. 2767–2772 (2009)
20. Tesch, M., et al.: Parameterized and Scripted Gaits for Modular Snake Robots. Advanced Robotics 23(9), 1131–1158(28) (2009)
21. Fu, Z., Zhao, Y., Qian, Z., Cao, Q.: Wall-climbing Robot Path Planning for Testing Cylindrical Oilcan Weld Based on Voronoi Diagram. In: Proceedings of the IEEE/RSJ International Conference on Intelligent Robots and Systems, Beijing, China, October 9-15, pp. 2749–2753 (2006)
22. Autumn, K., et al.: Robotics in Scansorial Environments. In: SPIE Unmanned Ground Vehicle Technology VII (2005)
23. Modular Snake Robots project page, http://www-cgi.cs.cmu.edu/afs/cs.cmu.edu/Web/People/biorobotics/projects/modsnake/index.html
24. Chonnaparamutt, W., Kawasaki, H., Ueki, S., Murakami, S., Koganemaru, K.: Development of a timberjack-like pruning robot: Climbing experiment and fuzzy velocity control. In: ICROS-SICE International Joint Conference, Fukuoka, Japan, August 18-21, pp. 1195–1199 (2009)
25. Woody project page, http://www.sugano.mech.waseda.ac.jp/woody/
26. Seirei Industry's pruning robot, http://www.seirei.com/products/fore/ab232r/ab232r.html
27. Saltaren, R., Aracil, R., Sabater, J.M., Reinoso, O., Jimenez, L.M.: Modeling, Simulation and Conception of Parallel Climbing Robots for Construction and Service. In: Proceedings of 2nd International Workshop and Conference on Climbing and Walking Robots (CLAWAR), pp. 253–265 (September 1999)

References

28. Almonacid, M., Saltaren, R., Aracil, R., Reinoso, O.: Motion Planning of a Climbing Parallel Robot. IEEE Transactions on Robotics and Automation 19(3), 485–489 (2003)
29. TREPA project page,
    http://arvc.umh.es/proyectos/trepa/
    index.php?type=proy&dest=inicio&lang=en&idp=trepa&ficha=on
30. Jones, B.A., Walker, I.D.: Kinematics for Multisection Continuum Robots. IEEE Transaction on Robotics 22(1), 43–55 (2006)
31. Kotay, K., Rus, D.: The Inchworm Robot: A Multi-Functional System. Autonomous Robots 8(1), 53–69(17) (2000)
32. Jiang, Y., Wang, H., Fang, L.: Path Planning for Inchworm-like Robot Moving in Narrow Space. In: Proceedings of the Joint 48th IEEE Conference on Decision and Control and 28th Chinese Control Conference, Shanghai, P.R. China, December 16-18, pp. 5977–5984 (2009)
33. Lim, J., Park, H., Moon, S., Kim, B.: Pneumatic robot based on inchworm motion for small diameter pipe inspection. In: Proceedings of the IEEE International Conference on Robotics and Biomimetics, Sanya, China, December 15-18, pp. 330–335 (2007)
34. Gravagne, I.A., Walker, I.D.: Manipulability, Force, and Compliance Analysis for Planar Continuum Manipulators. IEEE Transactions on Robotics and Automation 18(3), 263–273 (2002)
35. Provancher, W.R., Clark, J.E., Geisler, B., Cutkosky, M.R.: Towards penetration-based clawed climbing. In: 7th International Conference on Climbing and Walking Robots and the Support Technologies for Mobile Machines, Madrid, Spain, September 22-24 (2004)
36. Immega, G., Antonelli, K.: The KSI tentacle manipulator. In: Proceedings of the IEEE International Conference on Robotics and Automation, Nagoya, Japan, May 21-27, vol. 3, pp. 3149–3154 (1995)
37. McMahan, W., et al.: Field Trials and Testing of the OctArm Continuum Manipulator. In: Proceedings of the IEEE International Conference on Robotics and Automation, Orlando, Florida, May 15-19, pp. 2336–2341 (2006)
38. Chen, G., Pham, M.T., Redarce, T.: Development and kinematic analysis of a siliconerubber bending tip for colonoscopy. In: Proceedings of the IEEE/RSJ International Conference on Intelligent Robots and Systems, Beijing, China, October 9-15, pp. 168–173 (2006)
39. Robinson, G., Davies, J.B.C.: Continuum Robots - A State of the Art. In: Proceedings of the IEEE International Conference on Robotics and Automation, Detroit, Michigan, vol. 4, pp. 2849–2854 (May 1999)
40. Xu, K., Simaan, N.: Actuation Compensation for Flexible Surgical Snake-like Robots with Redundant Remote Actuation. In: Proceedings of the IEEE International Conference on Robotics and Automation, Orlando, Florida, May 15-19, pp. 4148–4154 (2006)
41. Camarillo, D.B., Milne, C.F., Carlson, C.R., Zinn, M.R., Salisbury, J.K.: Mechanics Modeling of Tendon-Driven Continuum Manipulators. IEEE Transactions on Robotics 24(6), 1262–1273 (2008)
42. Walker, I.D., Carreras, C.: Extension versus Bending for Continuum Robots. International Journal of Advanced Robotic Systems 3(2), 171–178 (2006)
43. Li, Z., Canny, J.: Nonholonomic Motion Planning. Kluwer Academic Publishers (1993)
44. LaValle, S.M.: Planning Algorithms. Cambridge University Press (2006)
45. Lozano-Perez, T.: Spatial planning: A configuration space approach. IEEE Transactions on Computing C-32(2), 108–120 (1983)

46. H῁ahnel, D., Burgard, W., Thrun, S.: Learning compact 3D models of indoor and outdoor environments with a mobile robot. Robotics and Autonomous Systems 44(1), 15–27 (2003)
47. Monnin, D., Schneider, A.L., Christnacher, F., Lutz, Y.: A 3D Outdoor Scene Scanner Based on a Night-Vision Range-Gated Active Imaging System. In: Proceedings of the Third International Symposium on 3D Data Processing, Visualization, and Transmission, Chapel Hill, North Carolina, June 14-16, pp. 938–945 (2006)
48. Dreyfus, S.E., Law, A.M.: The art and theory of dynamic programming. Academic Press (1977)
49. Hirose, S., Tsutsumitake, H.: Disk Rover: A Wall-Climbing Robot using Permanent Magnet Disks. In: IEEE/RSJ International Conference on Intelligent Robots and Systems, Raleigh, North Carolina, July 7-10, vol. 3, pp. 2074–2079 (1992)
50. Longo, D., Muscato, G.: SCID - A non-actuated robot for walls exploration. In: Proceedings of the IEEE/ASME International Conference on Advanced Intelligent Mechatronics, Como, Italy, vol. 2, pp. 874–879 (2001)
51. Fu, Y., Li, Z., Yang, H., Wang, S.: Development of a wall climbing robot with wheel-leg hybrid locomotion mechanism. In: Proceedings of the IEEE International Conference on Robotics and Biomimetics, Sanya, China, December 15-18, pp. 1876–1881 (2007)
52. Pack, R.T., Christopher, J.L., Kawamura, K.: A Rubbertuator-based structure-climbing inspection robot. In: Proceedings of the IEEE International Conference on Robotics and Automation, Albuquerque, NM, USA, April 20-25, vol. 3, pp. 1869–1874 (1997)
53. Unver, O., Uneri, A., Aydemir, A., Sitti, M.: Geckobot: A gecko inspired climbing robot using elastomer adhesives. In: Proceedings of the IEEE International Conference on Robotics and Automation, Orlando, Florida, USA, May 15-19, pp. 2329–2335 (2006)
54. Wile, G.D., et al.: Screenbot: walking inverted using distributed inward gripping. In: Proceedings of the IEEE/RSJ International Conference on Intelligent Robots and Systems, Nice, France, September 22-26, pp. 1513–1518 (2008)
55. Jia, Y., Tian, J.: Surface Patch Reconstruction From "One-Dimensional" Tactile Data. IEEE Transactions on Automation Science and Engineering 7(2), 400–407 (2010)
56. Okamura, A.M., Curkosky, M.R.: Feature-guided exploration with a robotic finger. In: Proceedings of the IEEE International Conference on Robotics and Automation, vol. 1, pp. 589–596 (2001)
57. Schopfer, M., Ritter, H., Heidemann, G.: Acquisition and Application of a Tactile Database. In: Proceedings of the IEEE International Conference on Robotics and Automation, Roma, Italy, April 10-14, pp. 1517–1522 (2007)
58. Astley, H.C., Jayne, B.C.: Effects of perch diameter and incline on the kinematics, performance and modes of arboreal locomotion of corn snakes (Elaphe guttata). The Journal of Experimental Biology 210, 3862–3872 (2007)
59. Cartmill, M.: Pads and claws in arboreal locomotion. In: Jenkins Jr., F.A. (ed.) Primate Locomotion, pp. 45–83. Academic Press, New York (1974)
60. Cartmill, M.: The volar skin of primates: its frictional characteristics and their functional significance. Am. J. Phys. Anthropol. 50(4), 497–510 (1979)
61. Cartmill, M.: Climbing. In: Hildebrand, M., Bramble, D.M., Liem, K.F., Wake, D.B. (eds.) Functional Vertebrate Morphology, pp. 73–88. Belknap Press, Cambridge (1985)
62. Hunt, K.D., Cant, J.G.H., Gebo, D.L., Rose, M.D., Walker, S.E., Youlatos, D.: Standardized descriptions of primate locomotor and postural modes. Primates 37, 363–387 (1996)

References

63. Saenz, D., Collins, C.S., Conner, R.N.: A bark-shaving technique to deter rat snakes from climbing red-cockaded woodpecker cavity trees. Wildlife Society Bulletin 27(4) (Winter 1999)
64. Pavlova, G.A.: Effects of Serotonin, Dopamine and Ergometrine on Locomotion in the Pulmonate Mollusc Helix Lucorum. The Journal of Experimental Biology 204, 1625–1633 (2001)
65. Federle, W., Brainerd, E.L., McMahon, T.A., Holldobler, B.: Biomechanics of the movable pretarsal adhesive organ in ants and bees. Proceedings of the National Academy of Sciences 98(11), 6215–6220 (2001)
66. Gray, J.: The Mechanism of Locomotion in Snakes. Journal of Experimental Biology 23(2), 101–120 (1946)
67. Murphy, M.P., Kim, S., Sitti, M.: Enhanced Adhesion by Gecko-Inspired Hierarchical Fibrillar Adhesives. ACS Appl. Mater. Interfaces 1(4), 849–855 (2009)
68. Zhang, X.J., Liu, Y., Liu, Y.H., Ahmed, S.I.-U.: Controllable and switchable capillary adhesion mechanism for bio-adhesive pads: Effect of micro patterns. Chinese Science Bulletin 54(10), 1648–1654 (2009)
69. Aksak, B., Murphy, M.P., Sitti, M.: Gecko Inspired Micro-Fibrillar Adhesives for Wall Climbing Robots on Micro/Nanoscale Rough Surfaces. In: Proceedings of the IEEE International Conference on Robotics and Automation, Pasadena, CA, USA, May 19-23, pp. 3058–3063 (2008)
70. Silva, M.F., Machado, J., Tar, J.K.: A Survey of Technologies for Climbing Robots Adhesion to Surface. In: Proceedings of the IEEE International Conference on Computational Cybernetics, Stara Lesna, Slovakia, November 27-29, pp. 127–132 (2008)
71. Scherge, M., Gorb, S.N.: Biological Micro-and Nanotribology, pp. 107–110. Springer, Heidelberg (2001)
72. Kim, K., Kim, Y., Kim, D.: Adhesion characteristics of the snail foot under various surface conditions. International Journal of Precision Engineering and Manufacturing 11(4), 623–628 (2010)
73. Jiao, Y., Gorb, S., Scherge, M.: Adhesion Measured on the Attachment Pads of Tettigonia Viridissima (Orthoptera, Insecta). The Journal of Experimental Biology 203(12), 1887–1895 (2000)

Index

2D arc fitting, 96, 124
2D manifold, 3, 117
3D arc, 94
3D arc fitting, 95, 124
3D path tracking, 4

capillary adhesion, 17

AB232R, 6
AB351R, 6
adaptability, 14
adhesive force, 57
admissible gripping position, 86
angle of twist, 86
approximated radius, 102
arboreal animals, 2
arboreal habitats, 9
artificial tree-climbing, 14
autonomous climbing, 2
autonomous tree climbing, 91

backdrive, 6
bark, 69
bending, 18
bending direction, 73
bio-inspired, 7
biomimetic pad, 18
bipedal locomotion, 10, 34

camera, 40
capillary adhesion, 9
caterpillar, 9
centerline, 73
central axis, 67

circumference, 86
claw penetration, 10
claw-gripping, 10
climbable workspace, 80
climbing gait, 34, 102
Climbing robot, 1
closer-half, 55
complexity, 14
compliance, 29, 30
concave triangular cone, 33
configuration, 73
configuration space, 117
continuum body, 27
continuum manipulator, 27
control architecture, 41
controllability, 55
convex triangular cone, 33
curvature of a tree, 69
cylindrical shape, 57
cylindrical structure, 7

decision-making, 43
deformation, 86
degrade, 18
design principle, 14
direction vector, 59
directional force, 66
dry adhesion, 9, 19
dynamic force, 17
dynamic programming, 117

elastomeric adhesive, 26
electroadhesion, 26
encoder, 34

equilibrium equation, 68
exploration, 92
exploring motion, 92
exploring strategy, 92
extend-contract, 15
extendibility, 29

farer-half, 55
fastening method, 9
feature-tracing, 92
fibrillar adhesion, 26
fitting quality, 127
flying, 10
foreword, VII
free body diagram, 68
frictional force, 10
front gripper, 34
front gripper frame, 76
front-gripper-based method, 123

Gough-Stewart platform, 1, 7
gripping angle, 10, 55, 57
gripping curvature, 57
gripping force, 10
gripping orientation, 60
ground station, 41
growth path, 92

helical shape, 11
hexagonal ring, 7
hopping, 10
hyper redundant, 15

inchworm, 9
inchworm locomotion, 10
inclined angle, 102
input coordinates, 73
insert angle, 57
inspection, 7
installation angle, 57
interlock, 10, 17

kinematic model, 73

least square method, 96
linear motor, 26
locking mechanism, 33
locomotion, 34
lumberjacks, 6

magnetic attraction, 26
maneuverability, 14
maneuvering method, 10
manmade structures, 1
master-slave, 41
maximum climbing slope, 84
mechanical spring, 30
metallic structure, 7

nonholonomic, 102

OCTARM V, 29

passive preloaded, 6
passive revolute joint, 67
path segmentation, 127
payload, 50
pedal wave locomotion, 9
penetration, 57
phalanx, 26
photovoltaic module, 40
pitch-back moment, 17, 30
pitching backwards, 14
plane fitting, 95, 124
pneumatic-driving, 27
posture, 73
pre-compressed, 67
preload force, 40
principal axis, 26
pull-in force, 60, 69
pull-out force, 69
pulse climbing, 10
pulvilli, 9

quadrupedal mechanism, 8

rack and pinion, 30
radar chart, 19
rank, 15
reachable workspace, 80
rear gripper, 37
rear gripper frame, 76
reconnaissance, 7
rescue, 7
RiSE V1, 7
RiSE V2, 2, 7
RiSE V3, 2, 8
robustness, 14

Index

self-contained module, 30
semi-passive Joint, 32
semi-passive joint, 85
shape reconstruction, 92
shear force, 57
side-toppling force, 17
six legs, 2, 7
size, 14
snail, 9
snake-like robot, 2, 8
speed, 15
squirrel, 10
state space, 118
suction, 9
surface curvature, 70
surveillance, 7
symmetrical footfall patter, 10
synchronize, 18

tactile sensor, 34, 91
target position frame, 80, 103
tendon, 73
Tentacle, 26
threaded rod, 5
three-dimensional space, 18
tilting sensor, 34, 100
toppling sideways, 14
tree data structure, 118
tree frame, 78, 98

Tree Gripper, 26
tree model, 98
Treebot, 24
Treebot-Auto, 37
TREPA, 7
tripod gait, 9
tripod pattern, 8
twisting angle, 33
two-bar linkage, 67

ultrasonic sensors, 7
unlocked state, 33
upper apex, 92

vacuum suction, 26
Van der Waals force, 9, 17
virtual tendon, 73
vision sensor, 91

wave length, 19
wet adhesion, 9
wheel-driven, 6, 15
wheel-driven method, 11
wire-driving, 27
WOODY, 1, 5
workspace, 2
worm gear, 6
wrap, 8

zero energy consumption, 26